The US
Media in the 20th Century

Robert T. Davis II

Occasional Paper 31

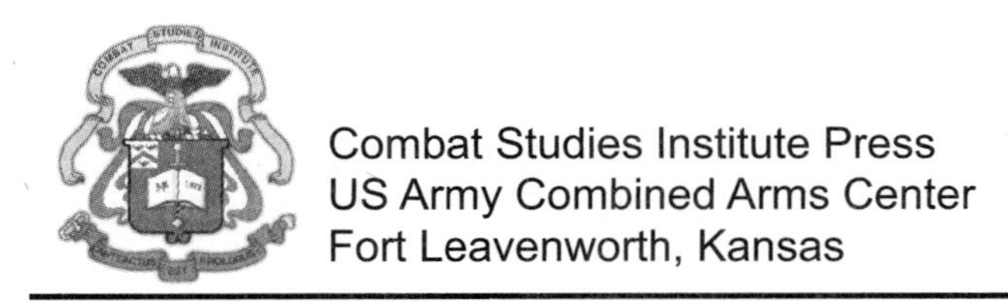

Combat Studies Institute Press
US Army Combined Arms Center
Fort Leavenworth, Kansas

Library of Congress Cataloging-in-Publication Data

Davis, Robert T., 1976-
The US Army and the media in the 20th century / Robert T. Davis II.
p. cm. -- (Occasional paper 31 / US Army Combined Arms Center)
Includes bibliographical references.
1. United States. Army--Public relations--History--20th century. 2. United States. Army--Press coverage--History--20th century. 3. Armed Forces and mass media--United States. 4. United States--History, Military--20th century. I. U.S. Army Combined Arms Center. II. Title. III. Title: U.S. Army and the media in the 20th century. IV. Title: United States Army and the media in the 20th century.

UH703.D385 2009
659.2'9355009730904--dc22

2009020829

First Printing: July 2009

CSI Press publications cover a variety of military history topics. The views expressed in this CSI Press publication are those of the author(s) and not necessarily those of the Department of the Army, or the Department of Defense. A full list of CSI Press publications, many of them available for downloading, can be found at: http://usacac.army.mil/CAC2/CSI/RandPTeam.asp.

Foreword

The Combat Studies Institute is pleased to present Occasional Paper 31, *The US Army and the Media in the 20th Century,* by Dr. Robert T. Davis II. Dr. Davis surveys the US Army's approach to media relations from the Spanish-American War to the first Gulf War. The relationship between the Army and the media is considered in the broader context of the US Government's approach to information management. Given the growing importance of information operations in 21st century warfare, this study provides a succinct overview of how the US Army has approached its relations with the media over the previous century.

The study highlights the recurrent tension that exists in both the Army and the US Government's information management writ large. This tension arises from the need for operational security and effective deception and psychological operations and the need to provide transparency to secure public acceptance and support for military operations. The long-running debate over how the Government's information management should be organized and operated reflects this tension. Thus, since World War I a number of bureaucratic manifestations of information management have been tried in wartime, including the Committee on Public Information, the Office of War Information, the Psychological Strategy Board, the United States Information Agency, and, most recently, the Office of Global Communications. With the exception of the United States Information Agency, whose tenure spanned the period from 1953 to 1999, all the other manifestations of bureaucratic information management rose and fell during the wars in which they were created. The growing pains of these organizations sometimes colored the Army's relationship with the media.

The need for units in the field to participate in information management is a major challenge for future operations. This study reminds us that those commanders who have gone out of their way to engage the media have, in many cases, had the greatest success with information management. *CSI—The Past Is Prologue!*

Dr. William G. Robertson
Director, Combat Studies Institute

Acknowledgments

I wish to thank my fiancé, Becky Miller, for tracking down a number of useful books and editing an early draft of this manuscript. The staffs of the Truman and Eisenhower Libraries were generous of their time and resources. I appreciate the efforts of my editor, Ms. Elizabeth Weigand, in preparing my manuscript for publication. A special thank you to Dr. William G. Robertson, Combat Studies Institute Director and Combined Arms Center Command Historian; former CSI Director Colonel Timothy R. Reese; and Mr. Kendall D. Gott, Supervisory Historian, Research and Publication Team, for allowing me to see this project through to completion.

Contents

Page

Figures

Chapter 1

Introduction

The area of freedom contracts and enforcement of restraints increases as the stresses on the stability of government and of the structure of society increase.

Frederick S. Siebert[1]

Cynics, I know, describe "PR" as a maternity gown designed to hide the true figure of fact. Undoubtedly, as abused by those who cover up or mislead, public relations can be stigmatized as mere propaganda or outright mendacity. Properly practiced, however, some form of it is necessary in a republic where the citizens must know the truth.

Dwight D. Eisenhower[2]

This study surveys how the US Army communicated its missions to the American public during periods of conflict within the context of the national policy toward information management. The phrase "information management" is used to suggest a host of interrelated terms to include censorship, information operations, information warfare, propaganda, public affairs, public information, psychological operations, psychological warfare, and strategic communications. The Army has always found its contribution to the US Government's information management policies challenging. With operational security a key objective of military planning, military commanders strive to deny information to the enemy. Information management, broadly speaking, serves two purposes. One purpose is to effect domestic morale. The second purpose is to undermine the enemy or enemies' morale and operational effectiveness. The proper relationship between these two functions has generated a long-running tension regarding the proper nature of the US Government writ large and, more specifically, the Army with regard to information management. On several occasions, specific practices of information management have blurred the distinctions between its two functions, which frequently leads to a domestic backlash against those information management structures and practices that develop in periods of tension. In almost all cases, it has been the preference of Government and military authorities to stress the importance of information management. At the same time, as Abraham Lincoln rightly noted in one of his celebrated debates with Stephen Douglas,

"Public sentiment is everything. With public sentiment, nothing can fail; without it nothing can succeed."[3] In the contemporary era, the traditional institutional interest in information management has had to contend with the growth and proliferation of modes of information dissemination. New media technologies have exacerbated the classic tension between the military's natural tendency to maximize the management of information to enhance security and effect the enemy versus the American public's desire for information. This study focuses primarily on one element of the broader theme of information management—the relationship between the US Army and the media in the period since the Spanish-American War.

The process of developing a formal information management capacity within the US defense establishment has taken place unevenly over time. In the 19th and early 20th centuries, ad hoc responses in times of crises were the norm. During World War I and World War II, more formal mechanisms were developed, but in many cases these responses did not survive in the postwar period. After American entry into World War I, the Committee on Public Information, better known as the Creel Committee, combined aspects of censorship and propaganda into a single office. Although the Secretaries of State, War, and Navy were all nominally represented in the committee's executive body, in reality it was George Creel, the director, who dominated the committee's outlook. The Creel Committee's perceived excesses and the poor relations between Creel and Congress led to the committee's rapid dissolution at war's end. Nevertheless, its legacy cast a long shadow over successive presidents' approaches to the control of information.

Though there was intermittent discussion of the need for military censorship in the years leading up to World War II, again, no formal mechanism was created until after America's entry into the war. President Franklin D. Roosevelt, aware of the legacy of the Creel Committee, never supported the creation of a single information agency. Instead, Roosevelt would ultimately countenance the creation of an Office of Censorship and an Office of War Information. The former deliberately eschewed the propagandistic characteristics of the Creel Committee. Like the Committee on Public Information, the Office of Censorship and Office of War Information were placed under civilian direction. And, like the Creel Committee, both were dissolved shortly after the end of the war (in both cases, by Executive order of President Harry S. Truman).

In the aftermath of World War II, the armed services retained public affairs roles. These roles were important tools of the military occupations in Germany, Japan, Austria, and Italy; moreover, they also retained the function of communicating with the American public. The rapid demobilization of the Armed Forces showed many leaders the need to educate the public about the continued requirements for national security. In this climate, the US Army established an Army Information School at Carlisle Barracks, Pennsylvania. During the heated controversy over roles and missions that characterized the scramble for appropriations in the late 1940s, there was a short-lived attempt by Secretary of Defense James V. Forrestal to create a single Office of Public Information in the Department of Defense (DOD). As was the case with many facets of defense unification, however, the Office of Public Information did not live up to Forrestal's original hopes, and the individual Services retained considerable autonomy in the direction of their public affairs efforts for many years.

With the outbreak of the Korean War, military public relations entered an uncertain era. The limited wars fought by the US military during the Cold War took place in more ambiguous settings than the "good war." These wars would challenge the ability of Army information management to fulfill its mission in often-difficult circumstances. The interest in strategic communications during the Global War on Terrorism, which included an attempt to establish an Office of Global Communications in the White House, demonstrated a very traditional (à la 20th century) attempt at information management. Its antecedents were found in the Creel Committee, the Office of War Information, and the Psychological Strategy Board. In addition, improvements in communications technologies would further complicate the issue. From television to cable to the internet, communications technologies have continued to increase the ability and speed to transmit information from the battlefield to the public, which defies the military to continue to shape the information environment of modern warfare.

This study analyzes the challenges, means, and practices that have constituted the US Army's response to public relations from the Spanish-American War to the Global War on Terrorism. It also alludes to the degree to which the Army's approach to public relations and public diplomacy contribute to strategic communications, as it is now understood. In this vein, US Army public relations are placed within the wider context of the US Government's evolving policies toward information activities. This study does not deal directly with psychological operations, deception operations, or electronic warfare; however, it does touch on past methods of media coverage, to include discussions of censorship, the use of reporters' pools, preferential access to senior commanders, and the practice of embedded reporters.

Notes

1. Quoted in Betty Houchin Winfield, *FDR and the News Media* (Urbana, IL: University of Illinois Press, 1990), 5.

2. Dwight David Eisenhower, *At Ease: Stories I Tell to Friends* (Garden City, NY: Doubleday & Company, 1967), 320.

3. Roy P. Basler, ed., "Lincoln-Douglas Debate at Ottawa" (21 August 1958), in *The Collected Works of Abraham Lincoln, Vol. III* (New Brunswick, NJ: Rutgers University Press, 1953), 27.

Chapter 2

Military-Media Relations through World War I

In time of peace, as well as war, editors and reporters carry secrets which they have no thought of publishing. They suppress items for editorial policy's sake, or which may do harm or unnecessarily hurt personal feelings, or in answer to an influential appeal over the telephone. There is censorship in the between-you-and-me and not-for-publication confidences of public men to a group of reporters; censorship in keeping the inside view of bruited gossip around a throne or the White House from the printed page; and in the whispered admonition "But the big boss wants it kept quiet" in any human organization. . . . Censorship is in human nature, its warrant and excuse in protectiveness. . . . In war censorship becomes official, direct, personified, war being itself often the product of the subtle underground censorship of peace.

Frederick Palmer[1]

The control of information in theaters of war has always been of central importance to military authorities. The need to control information in wartime was reflected in early legislation pertaining to the US Army. Under Article 57 of the 1806 Articles of War (in force until 1874), "Whosoever shall be convicted of holding correspondence with or giving intelligence to the enemy either directly or indirectly, shall suffer death or such other punishment as shall be ordered by the sentence of a court martial."[2] After the Battle of New Orleans, General Andrew Jackson briefly imposed martial law on the city and instituted press censorship. When a newspaper complained about Jackson's failure to lift martial law after the signing of the Treaty of Ghent, Jackson had the author of the article arrested for mutiny.[3]

The Civil War was the manifestation of deep tensions and contradictions that had existed in the United States since before the founding of the Republic. It, as many of the wars of the 20th century, had profoundly ideological characteristics. Both the armed forces of the United States and those of the Confederacy fought, at least in part, to defend their conceptions of ideal political association. President Abraham Lincoln clearly understood the deeply ideological nature of this conflict, and employed his own powerful rhetoric to advance and defend the cause of national union throughout the war. He served as propagandist-in-chief for the United States cause.[4] Within 4 months of the beginning of the Civil War, Lincoln's first Secretary of War, Simon Cameron, issued an Executive order that stated:

> All correspondence and communication, verbally or by writing, printing, or telegraphing, respecting operations of the Army or military movements on land or water, or respecting the troops, camps, arsenals, intrenchments, or military affairs within the several military districts, by which intelligence shall be, directly or indirectly, given to the enemy, without the authority and sanction of the major-general in command, be, and the same are absolutely prohibited, and from and after the date of this order persons violating the same will be proceeded against under the fifty-seventh article of war.[5]

When this first attempt at information control proved insufficient, a second Executive order was promulgated by Cameron's successor, Edwin Stanton, the following February. This subsequent order forbade any telegraphic communications relating to military operations that had not been previously authorized by the War Department or generals commanding armies in the field. Any newspaper that violated this order would subsequently be denied receipt of information transmitted by telegraph and prohibited from using railways for distribution.[6] Due to Stanton's influence, Postmaster-General Montgomery Blair extended this injunction by threatening to distribute any newspapers through the mail that violated the order. In the case of the Civil War, the limitations of communications facilities made this order practical. Stanton appointed Edward Sanford, president of the American Telegraph Company, as military supervisor of telegrams. To assure tight control over the telegraphic traffic, Stanton had the central telegraphic office moved from General George McClellan's headquarters to the War Department in a room directly adjoining his own office.[7] Control over the telegraphs provided no definitive resolution to military-media relations during the Civil War, however. Members of the press always retained the option of carrying their stories themselves or sending dispatches by courier. Though slower than being sent by telegraph, the advent of steamships and rail links meant that news could still travel faster than in earlier ages.

Perhaps no military officer in the Civil War typified the stress on military-media relations in wartime as General William Tecumseh

Sherman. Sherman was a complex man whose concern over failure often drove him into deep depressions. His fear of failure also drove him to control the situation in the field as closely as possible, which led him to the conclusion that there was little room for a critical press second-guessing his decisions. For Sherman, if war were to be waged successfully, it would be best conducted if the press were not present at all.[8] Sherman commanded a brigade at the first battle of Bull Run, where raw Union forces suffered an early reverse after having been urged "Forward to Richmond!" by Horace Greeley's *New York Tribune* and the general weight of Northern public opinion.[9] This no doubt contributed to his sense that the press's contributions to the war were less than helpful.

In the fall of 1861, Sherman was transferred to the Department of the Cumberland to serve under General Robert Anderson. There Sherman was charged with helping hold the key border state of Kentucky. While inspecting the fortifications on Muldraugh Hill, which protected the landward approaches to Louisville, Kentucky, Sherman was approached by F.B. Plympton of the *Cincinnati Commercial*. Plympton was interested in reporting on Sherman's theater and had gone to the trouble of bringing along several letters of introduction. These letters included one from Sherman's brother-in-law, Thomas Ewing Jr. After glancing at the letters of introduction, Sherman abruptly told Plympton to make sure he had himself on the next train back to Louisville. When Plympton protested that he was only there to convey the truth of what he saw to the public, Sherman dismissed him saying, "We do not want the truth about things; that is what we don't want. Truth, eh? No sir. You take that train to Louisville; we do not want the enemy any better informed about what is going on here than he is."[10] As the war progressed, Sherman remained hostile to the presence of reporters in areas under his command.

Once Union forces had captured New Orleans (April 1862) and Memphis (June 1862), Vicksburg, Mississippi, became the last major southern redoubt on the Mississippi River. With its commanding position on a bluff high above the river, Vicksburg's capture proved to be a considerable challenge. In early December 1862, Generals Ulysses S. Grant and Sherman met in Oxford, Mississippi, to plan a three-pronged assault on the city. To help maintain operational security, Sherman issued General Order Number 8, which prohibited any civilians, other than the transport crews, from accompanying his forces.[11] Due to complications, the initial three-pronged assault failed to materialize, and Sherman's troops were repulsed when they advanced toward the city on 29 December. However, subsequent press reports—ignorant that Sherman's component was only one part of the intended three-pronged attack and eager to vilify Sherman for his policies toward the press—took advantage of the failure to resurrect charges about Sherman's mental competence for duty. The abusive press criticism grew when it was announced that Sherman was being replaced as commander by Major General John McClernand, who, in fact, had been appointed by Lincoln to mollify public opinion in McClernand's home state of Illinois.[12]

Sherman responded by court-martialing Thomas W. Knox, a correspondent for the *New York Herald* and author of a scathing critique of Sherman's initial assault on Vicksburg.[13] Though Knox had violated General Order Number 8, providing Sherman with a pretext, John Marszalek argues that it was the principle of press exclusion rather than punishing an individual that interested Sherman.[14] Sherman wanted to establish the precedent that civilians accompanying military forces were subject to military law.[15] Knox was charged with giving information to the enemy (see Cameron's August 1861 order above), being a spy, and disobeying orders. Knox's counsel, Lieutenant Colonel W.B. Woods, countered these three charges with information provided by Knox. As to the first charge, Woods was able to argue that General McClellan had modified the initial Cameron order, allowing the press to write about units and give commanders names so long as the accounts were written after the fighting had ended. Knox also produced at his trial a pass from General Grant. Finally, Woods argued that Knox had been unaware of General Order Number 8 until after he was on the transport accompanying Sherman's force. The court-martial found Knox not guilty on the first two charges. He was found guilty of disobeying General Order Number 8, but the court-martial attached "no criminality thereto." As a result of the decision, Knox was to leave Army lines under penalty of arrest should he return.[16]

Though Sherman was unhappy that the court-martial had seen fit to find Knox not guilty on two of the charges and guilty of the third on a technicality, in the end the victory was Sherman's. Other national newspapers did not rally to Knox's defense. Two reporters, Albert Richardson of the *New York Tribune* and James Mitchell of the *New York Times*, did plead Knox's case to President Lincoln. Lincoln skillfully deflected their appeal by agreeing to write a letter under which Knox could proceed to General Grant's camp, but the final decision on whether Knox could remain in the Army's theater of activity would reside with Grant. Grant sided with his subordinate, and Knox moved back to covering the war in the east. Because of this famous court-martial of a reporter, which has never again been attempted by military authorities, General Sherman helped establish a precedent for the accreditation of reporters subject to the approval of theater commanders.[17]

Spanish-American War

The Spanish-American War (1898) marks an exceptional point in the history of military-media relations. The jingoistic mood that characterized the populace at large and much of the press created a situation in which the reporters' enthusiasm for war reporting eclipsed any need for the military to attempt to shape the public's perception of the war.[18] This was, after all, the age of the muckraker and yellow journalism. The brevity of the conflict and its successful outcome were certainly contributory factors to the generally positive coverage of the war's actual combat phases. Nonetheless, even the Spanish-American War was attended by its own information management difficulties. Generally, the criticism that befell the Army was aimed at the quality of medical attention given to soldiers and the quality of supplies provided to the forces. Later assessments, however, have tended to mitigate these shortcomings given the haste with which the US armed forces where propelled into the war.

The public relations difficulties that followed the successful prosecution of the war had everything to do with the nature by which a small Territorial Army augmented by volunteers had been rushed off to war. Before the Spanish-American War, the US Army had consisted of 25,000 men scattered in small detachments across the country. By late 1899, the US Army stood at 100,000 men, with two-thirds of this force stationed in the Philippines, Guam, Cuba, and Puerto Rico.[19] That the war had been short and decisive was all the better, as the Army was institutionally ill prepared to support a large number of troops, either at home or abroad, for long. Indeed, after the conclusion of the war, a number of shortcomings in both logistical and medical preparations became apparent.[20] As was typical of 19th century military campaigns in the tropics, the outbreak of yellow fever and malaria among American troops in Cuba threatened to be deadlier than had combat with the Spanish.[21] The problem first became apparent among the Fifth Corps, under the command of Major General William R. Shafter, who had recently defeated Spanish forces in Santiago, Cuba.

On 3 August 1898, Shafter wired the War Department to inform them that 75 percent of the recently victorious Fifth Corps had come down with malaria and was now in actuality an "army of convalescents." Though Shafter had reported some earlier cases of yellow fever and malaria, throughout July his dispatches had indicated that his unit remained combat effective. The Army had made some recommendations to rotate sick troops home by establishing a rest and recreation camp, subsequently named Camp Wikoff, at Montauk Point on the eastern end of Long Island.

However, initial plans called for rotating small detachments of the unit through the camp. Preparations for receiving large numbers of troops, let alone the entire Fifth Corps, had simply not been prepared prior to Shafter's dispatch.[22]

Accompanying Shafter's dispatch were supporting letters from his division commanders and brigadiers that were generally quite blunt as to the conditions of their troops.[23] These materials, leaked to an Associated Press correspondent at Shafter's headquarters, appeared the following day on newspaper front pages around the country. The resulting public furor about supposed Army callousness unfortunately contributed to the decision to send all of Fifth Corps' troops to Camp Wikoff on the hopes that preparations could be completed before the troops arrived. This did not happen. Instead, the Army was left with a public relations disaster as sick troops arrived at a camp where the infrastructure and medical facilities were simply overwhelmed. With the camp located in Long Island, only a short train ride from New York, the sad condition of Fifth Corps was exposed to the full glare of public scrutiny. Of the 21,000 troops who passed through the camp, 257 died.[24]

The bad press that resulted from the travails of Fifth Corps was further exacerbated by the actions of Commanding General of the Army Nelson Miles. Miles publicly aired a number of his own grievances with Secretary of War Russell A. Alger and later charged that the Army had distributed tainted beef, with the implication of malfeasance. Miles opened the dispute with Alger publicly while he was leading the campaign in Puerto Rico.[25] He informed W.J. Whelply of the *Kansas City Star* on 11 August that Secretary Alger had undermined his authority, charged that the War Department had ignored his plans for handling the yellow fever epidemic, withheld ships and troops from his Puerto Rico expedition (some Fifth Corps forces had been earmarked to Miles as reinforcements before they were all ordered back to Montauk Point), and garbled his communiqués when releasing them to the press.[26]

As a result of the general criticism leveled against the Army, Secretary of War Alger formed a commission to investigate the conduct of the war. Grenville M. Dodge, a retired Civil War general, agreed to serve as head of a commission, known henceforth as the Dodge Commission. The Dodge Commission conducted extensive hearings from October through December.[27] When Miles testified in late December, he resurrected his public campaign against Alger and Commissary General Charles Eagan, who he implied were responsible for tainted beef, which was "one of the serious causes of so much sickness and distress on the part of the troops."[28]

Eagan would subsequently ruin his career in an impassioned defense of his actions, and ultimately the Dodge Commission would find the charges against Miles largely unfounded. In the end, the much publicized dispute between Alger and Miles did neither man much credit and only served to heighten domestic criticism of the Army. In the immediate term, the Dodge Commission's report accomplished little, but its findings would soon provide grist to the efforts of Alger's successor as Secretary of War Elihu Root, who would initiate wide-ranging reforms of the US Army.[29] However, before these reforms took shape, the Army became involved in a difficult war subduing Filipino insurgents who opposed US annexation of the territory. Rather strained military-media relations accompanied this episode.

Philippine-American War

The Spanish-American War in the Philippines ended with US forces in occupation of the capital of Manila, but with much of the interior under the control of the Filipino nationalist and miscellaneous regional forces who had been fighting an insurgency against the Spanish since 1896. In the 5 months after fighting against the Spanish ended, the US forces and Filipinos nervously eyed each other while the future of the Philippines was determined by negotiations in Paris. The US forces that had been dispatched by President William McKinley to the Philippines had been initially sent under vague instructions. When the Eighth Corps landed south of Manila on 25 July, it numbered approximately 12,000 men. The US Army forces had only rough parity with the Spanish forces in Manila, compensated by Commodore George Dewey's squadron in Manila Bay. After the Spanish had been defeated, however, the Eighth Corps faced a rather different challenge. The simmering tensions between Filipino nationalists, led by Emilio Aguinaldo, and the US troops who occupied the city broke into full-scale conflict on the night of 4 February 1899.[30] Though both sides would later claim the other was responsible for the outbreak of fighting, the most recent scholarship on the Philippine-American War points out that there is little indication either side was seeking the outbreak of hostilities at that time.[31] Once fighting broke out between Filipino nationalists and US forces, however, the shortage of US manpower, exacerbated by the toll of disease in tropical campaigns, severely impeded the subsequent conventional campaign. Nonetheless, within a year US forces would have control over most of the major towns of the Philippines. Defeating the Filipino forces proved a more difficult challenge, perhaps not surprisingly given the combination of geography (7,000 islands covering 500,000 square miles) and politics (Democrat-nominee William Jennings Bryan was running against President McKinley in 1900 on an anti-imperialist platform, which encouraged Filipino resistance until the results of the US election were known).[32]

Major General Elwell S. Otis was serving as military governor with command of the Eighth Corps at the time of the outbreak of hostilities.[33] Otis was a veteran of the Civil War, subsequently trained in law at Harvard University, who went on to a distinguished career in the US Army. This included having been selected by General Sherman to serve as the first commandant of the School of Application for Cavalry and Infantry at Fort Leavenworth, Kansas. Otis later served as commander of the Regular Army's 20th Infantry Regiment, commander of the Departments of Columbia and Colorado, and second-in-command of the original expedition to the Philippines, for which he was widely credited with the successful deployment of the Eighth Corps to the Philippines.[34] Despite his distinguished record and considerable organizational gifts, Otis's relation with the press was often strained. Even before the Eighth Corps departed San Francisco for the war with Spain, some of the press had developed a negative attitude about the future commander.[35] This should not suggest that military-media relations on the whole were not fairly good in the early phase of the Philippine-American War. Many of the correspondents in the Philippines shared the general imperialistic outlook of the public at large and reveled in the adventure of covering the Army in the field. Frederick Palmer, who would later serve as press censor to General John Pershing in World War I, would recollect that many of his fellow correspondents would participate in combat operations "out of sheer fellowship."[36] Clearly, objectivity was not yet the overriding concern that it would become to later generations of journalists.

When fighting broke out between the Filipino and US forces on 4–5 February 1899, General Otis was confident that American forces would prevail. On 6 February, the *Boston Daily Globe* carried an article that stated "every confidence is felt . . . that Gen. Otis is the master of the situation."[37] The relative ease with which US forces drove the Filipino forces out of the environs of Manila soon prompted Otis to assume that Filipino resistance would quickly be broken. This in turn encouraged his subsequently parsimonious requests for resources and troops. When the Filipino resistance proved to be more resilient than was initially thought, this parsimony became one point of criticism on which the press fastened. On 12 February, Otis cabled the War Department and informed them of his belief that if those reinforcements already dispatched (a force under General Henry W. Lawton was en route) were available, it would "probably end [the] war or all determined active opposition in twenty days."[38]

Otis initially foresaw a war against the insurgents that would be relatively brief and undemanding of manpower. The arrival of more Regular Army troops became an important issue, as the terms of service for the Volunteer units were tied to the official end of the Spanish-American War.[39] Otis's rosy view of the war at this stage may have been geared toward reassuring those Volunteers present that the fighting would quickly be over once the US forces had convinced the Filipinos that they were overmatched. In any case, overly enthusiastic reports soon appeared in US papers. On 18 March a *New York Times* front page headline proclaimed "CLIMAX AT HAND IN THE PHILIPPINES."[40] The article was purportedly based on the positive situation reports that Otis cabled to the War Department, which suggested that the Filipino resistance might suddenly collapse. In fact, it would be 3 more years before President Theodore Roosevelt would proclaim an end to the fighting in the Philippines, and even then there were sporadic campaigns until 1913. At the same time as the *New York Times* printed Otis's rather positive assessment of the situation, a more critical story, telegraphed from Hong Kong—beyond the reach of the US military censors in Manila—suggested a rather different story. Oscar Williams, a US consul, was quoted as saying that he "didn't expect to live to see the end of the war." The Filipino insurgents, rather than being on the verge of collapse, were said to be encouraged by the practice of the US forces retiring to their own lines after engagements.[41] This latter report was symptomatic of the growing divergence between Otis and the press correspondents' views of the war.

By mid-July, with no collapse in the Filipino resistance, the press's comments about Otis's leadership were turning more critical. Reporter James Creelman charged, "Unless Gen. Otis is removed and a competent general put in command, the whole campaign will be a failure."[42] Increasingly, the press corps in Manila (never more than a score of correspondents) found cause for complaint in Otis's censor. After two meetings between Otis and the press corps failed to rectify the situation, the members of the press corps again went around Otis's censor by dispatching a note to Hong Kong for transmission back to the United States. On 18 July, a number of papers in the United States carried the letter that denounced Otis's censorship policy and the way he portrayed the war in his official reports.[43] When pressed for his version, Otis argued that the press correspondents had been denied permission to cable stories that claimed "official reports sent misrepresented conditions." Otis argued that his own reports contained no deliberate misrepresentations, and sometimes bordered on being "too conservative." When he had pushed the press corps to offer specific instances in which he had misrepresented the situation on the ground, they countered by arguing that it was not the situation reports but rather Otis's conclusions they felt were unwarranted.[44] Despite the criticism leveled against General Otis in the press, President McKinley never wavered in his support of Otis.[45] Despite the ongoing hostility of the press corps, Otis would remain in command until 5 May 1900, stepping down at his own request when he felt confident that the guerrillas were nearly defeated.[46] However, it would take another 2 years, under the direction successively of Major General Arthur MacArthur and Major General Adna Chaffee, before the Filipino resistance would subside.

Though small-unit actions had been an element of the fighting in the Philippine-American War from its beginning, Filipino forces did not formally resort to guerrilla warfare until late November 1899.[47] It is this phase of the Philippine-American War that has been the subject of the greatest disapprobation. Allegations of atrocities and torture by American soldiers—Filipino forces carried out similar activities—dogged the US military during the campaigns in the Philippines. The conflict between Commanding General Miles, President Roosevelt, and Secretary Root exacerbated public criticism of the American occupation. The summer of 1902 was the climax of a number of highly publicized courts-martial of US military figures in the Philippines.[48] In March 1902, news broke that two Marines, Major Littleton W.T. Waller and Lieutenant H.A. Day, were court-martialed for executing natives on the island of Samar without trial. Eventually, both Waller and Day would be acquitted on the grounds that they were following orders from their superior officer, US Army Brigadier General Jacob H. Smith, who infamously and perhaps apocryphally had given the order to turn Samar into a "howling wilderness."[49] These courts-martial contributed to the widespread disgrace of the Army's conduct of the counterinsurgency in the Philippines, which was only partially curbed through the characteristically energetic involvement of President Roosevelt. As Brian Linn has argued, "Samar cast a pall on the army's achievement and, for generations, has been associated in the public mind as typifying the Philippine War."[50]

The Army's reputation was also ill served by the increasing public dispute between Commanding General Miles and Secretary Root. In December 1901, Miles made public comments on a decision reached in a US Navy Court of Inquiry. Both Roosevelt and Root were frustrated that the Commanding General of the Army would insert himself in the affairs of another Service, and Roosevelt became sufficiently aroused to instruct Root to formally reprimand Miles.[51] Miles, though nominally a Republican, was not without political ambitions, and with Roosevelt increasingly looking to garner the Republican nomination in 1904, began to position himself to attract the vote of the anti-imperialist faction of the Democratic Party. In the wake of Roosevelt and Root's reprimand, Miles found an opportunity to advance his own political ambitions by embarrassing the administration over the question of atrocities in the Philippines. In mid-March, Miles granted an interview to Henry Watterson, whose account was published in the *Washington Post*. It reported that the administration had denied Miles permission to conduct an inquiry into the situation in the Philippines, hinting that reports of Army misconduct were being suppressed.[52] Later that month, Miles testified before the Senate Military Affairs Committee against Secretary Root's proposed Army reform to establish a General Staff, calling such a measure subversive to the Army's interests and threatening resignation if it were enacted.[53] Finally, Miles was responsible for passing on the report of Major Cornelius Gardener, the military governor of Tayabas province in the Philippines, to Roosevelt's critics in the Senate. The report, which Roosevelt supporter Senator Henry Cabot Lodge reluctantly allowed to be published on 11 April, spoke of the brutalization of the province inhabitants by American soldiers who had generated "deep hatred" toward the American occupiers as a result.[54]

Ultimately, President Roosevelt himself had to organize a campaign in defense of the Army's reputation. Roosevelt requested that Lodge provide the Senate with ample evidence of Filipino atrocities against American troops. The President also saw to it that General Smith was dismissed from the Army, though the court-martial in Manila had only admonished him for excessive zeal when it returned its verdict in mid-July. Lodge's Senate defense, the decision to dismiss Smith, and President Roosevelt's declaration on 4 July 1902 that the Philippine "insurrection" was officially over did a good deal to mute the storm of criticism that the courts-martial and General Miles had stirred up over the previous months.[55] Though there was much discussion in the press about Miles being forced to resign, Roosevelt ultimately chose simply to announce that on Miles' 64th birthday he would be retired from Service. Despite a distinguished record in the Indian Wars of the latter 19th century and a career stretching back to the Civil War, Roosevelt and Root saw to it that Miles' retirement ceremony on 8 August 1903 was coldly formal and lacked any of the normal trappings that traditionally accompanied the retirement of senior generals.[56]

World War I

Before the United States entered World War I, there had been considerable discussion of the issue of censorship. American newspapers carried numerous stories regarding the censorship policies of the belligerent states after the outbreak of fighting in Europe. When President Woodrow Wilson decided to launch a punitive expedition into Mexico in response to Pancho Villa's attack on Columbus, New Mexico, the US Army imposed censorship on its field activities.[57] Drawing on this experience and the example of the European belligerents, the War College Division of the General Staff created a plan of censorship to be put in place should the United States be drawn into the European conflict.[58] Ultimately, President Wilson authorized the creation of a Committee on Public Information, which combined censorship and propaganda functions under civilian, rather than military, direction. The Committee on Public Information became increasing unpopular with congressional critics of the President, and was dissolved quickly after the end of the war. Revelations in the 1920s and 1930s about the extent to which belligerent governments attempted to manipulate public opinion through propaganda campaigns created a longstanding hostility among the American public toward any attempt to replicate the experience of the Creel Committee.

By early 1917, the Wilson administration was moving toward abandoning neutrality and declaring war on Germany. After months of searching for an appropriate response to the resumption of unrestricted submarine warfare by Germany, President Wilson requested that Congress approve a Declaration of War in his address to a joint session of Congress on 2 April. Wilson deployed his typically lofty rhetoric about the purpose of American entry into war. At the same time, he advanced several points that are more practical. Wilson called for the United States to advance "liberal financial credits" to those governments already at war with Germany; fully equip the Navy, with particular emphasis on dealing with Germany's submarines; and the call up of 500,000 men, "chosen on the principle of universal liability to service."[59] The President also made note in his address of the millions of US citizens of German birth and ancestry who lived in America, noting that the vast majority were "true and loyal Americans as if they had never known any other fealty or allegiance." But given the numerous intrigues by the German Government's agents in the US during the period of official United States neutrality, Wilson also threatened that "if there should be disloyalty, it will be dealt with with a firm hand of stern repression."[60] To better communicate the lofty ideals that Wilson attributed to America's entry into the war and to better ensure the broad support of the American people for the war effort, the President soon moved to establish an organization to coordinate and disseminate propaganda both at home and abroad.

On 12 April 1917, President Wilson tentatively approved the appointment of George Creel as head of a new Committee of Publicity. Creel himself envisioned a body which was to oversee the US Government's efforts at gaining public confidence and stimulating recruitment, while at the same time suppressing (Creel disliked the term censorship) information that the Government did not want in circulation.[61] After a meeting the following day, the Cabinet issued a press release announcing the creation of the Committee on Public Information (CPI). It was composed of Secretary of State Robert Lansing, Secretary of War Newton D. Baker, Secretary of the Navy Josephus Daniels, and George Creel who served as chairman.[62]

Before the ink was dry on the initial authorization, Secretary Lansing voiced his concern that the CPI not be perceived as superseding the functions of the State Department. Once the CPI was up and running, Lansing remained concerned that the CPI's publicity campaigns not undermine formal diplomatic communications with foreign governments.[63] In late June 1917, the Creel Committee would come under criticism from the Army. This occurred when news of the arrival of the first units of General John Pershing's expedition appeared in print before the Army desired it to be made public. Under the presumption that anything sent from France had already passed a censor, Creel's original guidelines to the press had expressly permitted the publication of news from received cables.[64] Secretary of War Baker briefly considered trying to assert War Department censorship over incoming cables from France (the Navy Department already had control of the telegraph stations), but the situation was cleared up by the use of a system of accreditation of reporters to Pershing's headquarters in France.[65] Despite some of the initial friction over the committee's direction, during the slightly less than 2 years of its operation, Creel built the CPI into a large-scale propaganda organ.[66] The Creel Committee was a chief source and the primary disseminator of US propaganda at home and abroad, while the US military role in public affairs was largely relegated to its handling of censorship in theater.

The attitudes of the Army Expeditionary Force (AEF) commander, General Pershing, did not greatly deviate from that of his fellow commanders in the British, French, or Belgian Armies. In his memoirs, Pershing wrote:

> The special purpose of censorship and other precautions to prevent the publication of military information was to keep the enemy from learning our plans and movements. Secrecy gives a commander the possibility of surprising his opponent and the surest road to defeat would be to let the enemy know all about one's preparations. The operations of an army cannot be successfully conducted under any such open methods. It was impressed upon our

> forces and upon the correspondents that every person who, either willfully or inadvertently, disclosed facts of military value thus gave the enemy an advantage, and that such person, if in the army, might actually be responsible for the unnecessary sacrifice of his own comrades.[67]

At the same time, Pershing recognized, perhaps belatedly, that "the suppression of news prevented our people from obtaining a clear and contemporaneous conception of the great and oftentimes brilliant achievements of our armies."[68] Pershing's chief censor and press officer in World War I was Frederick Palmer.[69] Palmer, an experienced journalist covering worldwide military campaigns, started as a military correspondent in the Greco-Turkish War of 1897.[70] He went on to cover the Philippine-American War, the Boxer Rebellion, the Russo-Japanese War, and the Balkan Wars. In 1915, he became the lone American correspondent accredited by the British Government to cover fighting on the Western Front. Once the United States entered the war, the *New York Herald* offered Palmer a lucrative contract to cover the war for $40,000. Palmer turned this down to don a major's uniform with the standard salary of $2,400 per year.[71] He was charged with supervising the accreditation of correspondents, who had to post a $1,000 bond subject to forfeiture if they violated any rules, including the right of the AEF G-2 section to censor their dispatches. Initially the press corps was limited to 31 members, but pressure from the media soon led to a dramatic expansion of the press corps. At its height, Palmer was responsible for overseeing as many as 411 correspondents.[72] Palmer recognized the selfless nature of the job, writing in his memoirs that "A censor can have no friends," because he was a "professional no-man." The propagandist, by contrast, was a "professional yes-man" whose business is to make friends.[73]

The propaganda effort the Creel Committee organized was reinforced by three wartime acts, which placed restraints on First Amendment rights. These were the Espionage Act, enacted on 15 June 1917; the Trading-with-the-Enemy Act, enacted on 6 October 1917; and the Sedition Act, enacted on 16 May 1918. The Espionage Act prohibited any type of news or report that interfered with US national security, military operations, recruitment, or incited mutiny or disloyalty. The Espionage Act gave the Postmaster-General authority to ban from the mail any publication that did any of the above. The Sedition Act attempted to clarify ambiguities of the Espionage Act by spelling out specific speech offenses against the US Government, which were prohibited until the act's repeal in 1921.[74] The Espionage Act, which was upheld by the Supreme Court in *Schenck vs.*

United States (1919), remains in effect to this day. The Justice Department prosecuted over 2,000 Americans under the provisions of the two acts during the war, securing around 1,000 convictions.[75]

The CPI was organized into a domestic division and a foreign division. The domestic division was charged with helping to mobilize the population of the United States to support the war. One element of the domestic operation, known as the Foreign Language Newspaper Division, was staffed with 200 volunteers who watched foreign-language newspapers published in the United States for violations of the Trading with the Enemy Act.[76] The foreign division was split into a Foreign Press Bureau, the Wireless and Cable Service, and the Foreign Film Service. CPI's foreign division established offices in over 30 countries.[77]

In an era before radio had fully come of age, the CPI was primarily reliant on the printed and spoken word to spread its message, supplemented by media such as the weekly newsreel, which was then very popular in American movie houses. World War I was one of the first wars that employed the technology of motion pictures. Almost no footage of the war ever made its way to American audiences, as the belligerents instituted a widespread ban on filming by civilians. The US Army Signal Corps was responsible for most of the footage that did make its way into American newsreels. Much of the footage shot by the Signal Corps, however, was never seen. Many of the images of the war front were considered too stark for home-front audiences and were censored by the Creel Committee's Division of Pictures.[78] One of the few uses of Signal Corp footage was in the propaganda film *Pershing's Crusaders* (1918). The film opened with a picture of General Pershing flanked by two knights in medieval garb. The opening subtitles read, "The world conflict takes upon itself the nature of a Crusade. . . . We go forth in the same spirit in which the knights of old go forth to do battle with the Saracens." After this allusion, it transitioned to a message that resonated with the highly moralistic rhetoric of President Wilson, stating, "The young men of America are going out to rescue Civilization. They are going to fight for one definite thing, to save Democracy from death. They are marching on to give America's freedom to the oppressed multitudes of the earth."[79] The film, one of the few of its kind produced for American audiences, apparently did little to excite audiences.[80]

Given the heavy censorship of films and photographs, the vast bulk of the Creel Committee's effort was focused on the written and spoken word. To a large degree, the Creel Committee's approach was aimed at loading the American media down with facts that were perceived as

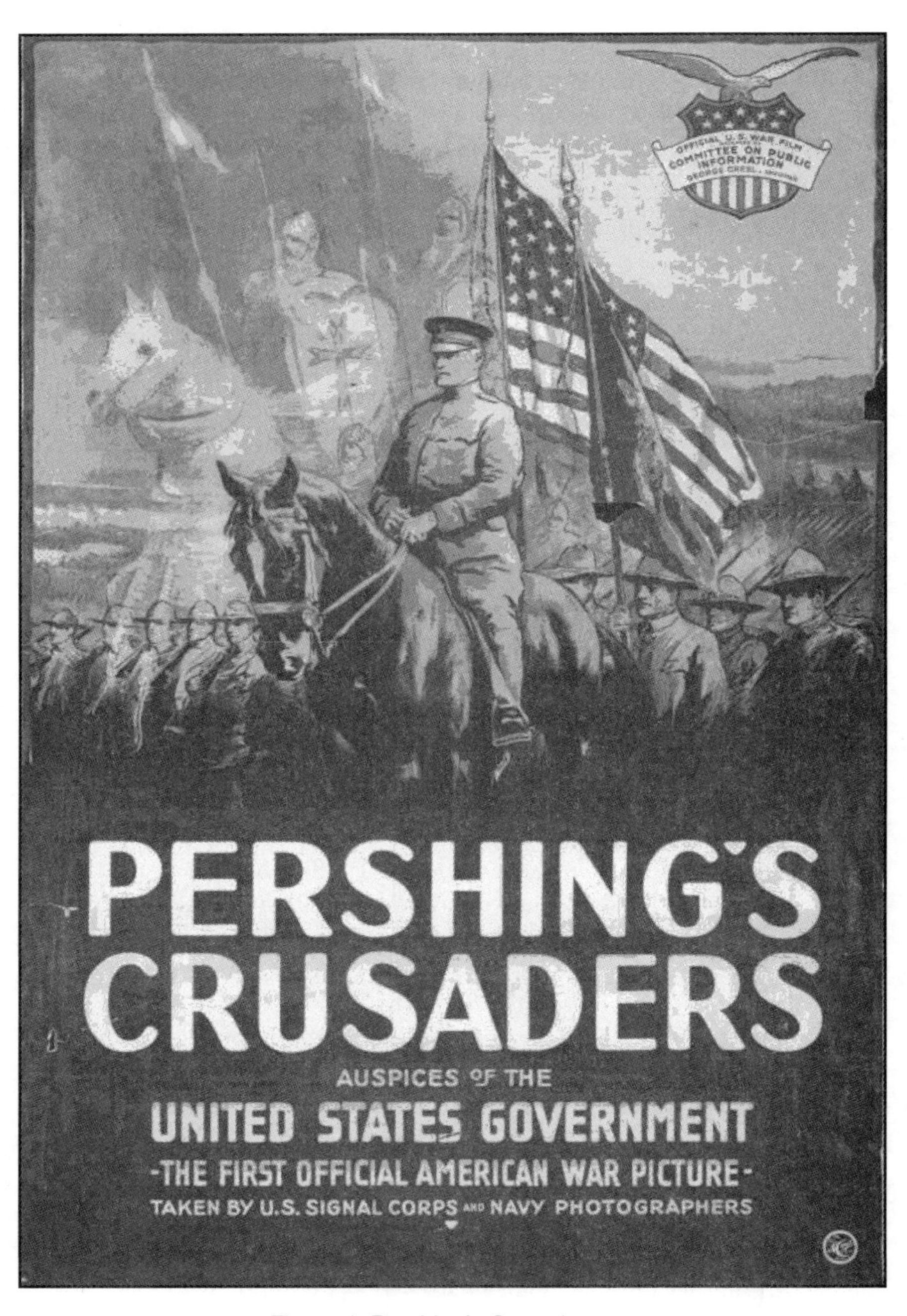

Figure 1. Pershing's Crusaders.

aiding the war effort. CPI's Division of News, the first to be set up and put into operation, was responsible for issuing press releases. Once it was up and running, the Division of News averaged 10 press releases a day. In addition, an official government paper, the *Official Bulletin*, was printed daily, except on Sundays, from 10 May 1917 through 31 March 1919. The *Official Bulletin* included presidential speeches and proclamations, information related to casualties, and orders issued by the Cabinet and other war-related agencies.[81] Daily circulation rose from 60,000 in May 1917 to 115,000 in October 1918. The *Official Bulletin* provided anywhere from 8 to 40 pages of material that the Creel Committee deemed of value.[82] Though the subscription rate was set deliberately high so the Government did not appear to be competing with private-media outlets for circulation, free copies were distributed to newspapers, Government officials, military bases, and posted in approximately 50,000 US Post Offices.[83]

The Creel Committee's Division of Civic and Educational Cooperation was responsible for producing over 100 pamphlets, which had a combined printing of 60 million copies. These pamphlets included titles such as *How the War Came to America* (5,428,000 copies), *Conquest and Kultur: Aims of the Germans in Their Own Words* (1,203,000 copies), *German War Plots and Intrigues in the United States during the Period of Our Neutrality* (127,000 copies), *American Interests in Popular Government Abroad* (597,000 copies), and *America's War Aims and Peace Terms* (719,000 copies).[84] One of the most successful CPI programs of the war, which developed into its own division, was the volunteer speakers, who became known as the Four Minute Men. These volunteer speakers delivered set speeches on a range of topics developed by the CPI and spelled out in the *Four Minute Men Bulletin*. The division was so-named as the speaker's were supposed to deliver their speeches in approximately 4 minutes. There were eventually Four Minute Men active in all 48 states and US territories. Generally, the Four Minute Men addressed audiences in movie theaters before the feature film, with the 4-minute time limit imposed to avoid overstaying the hospitality of theater owners. In May 1917, there were 1,500 speakers, with the number growing to 40,000 by September 1918. The Four Minute Men's speeches ranged over topics such as supporting bond drives, food conservation, justifying the federal income tax as necessary for the war effort, "Why Are We Fighting," and "Unmasking German Propaganda." George Creel estimated that the Four Minute Men gave a million speeches to a combined audience of 400 million listeners.[85] The wide-ranging activities of the CPI abruptly came to a close shortly after the end of the war when Congress moved quickly to dissolve the committee.

In the wake of World War I, the American public turned strongly against the internationalist, activist vision of America that President Wilson and the Creel Committee had so recently championed. Historian David Traxel has suggested that the widespread disillusionment with the "Great Crusade" was in part so common because Government-sponsored propaganda, in an effort to justify American participation in the war, had built up people's expectations for improvement in the world "out of all proportion to reality."[86] The Senate's failure to ratify the Treaty of Versailles, including rejection of US membership in the League of Nations, signaled a familiar inward turn of the American polity.[87] With the war's end, Pershing's crusaders would quickly demobilize. Between the Armistice in early November 1918 and the end of June 1919, 2,608,000 enlisted men and 128,000 officers accepted discharges from the Army. This left a mere 130,000 men under arms to supervise potential American occupation responsibilities in Germany and elsewhere.[88] Retrenchment in defense, a familiar pattern in the annals of the US Army, became the order of the day. From the Washington Naval Conference in 1921–22 to the ratification of the Kellog-Briand Pact by the US Senate in January 1929 (in which the United States and numerous other nations renounced war as an instrument of national policy), disarmament, not defense, was most prone to stir public interest.

When the Army was in the news in the interwar period, often as not it garnered little public good will. On two occasions during the interwar period, the Army found itself at the receiving end of rather negative publicity. The celebrated court-martial and subsequent conviction of air power enthusiast General Billy Mitchell in 1925 provided an example for many officers of the danger to one's career of too open an expression of personal views. The personal role of Chief of Staff Douglas MacArthur in breaking up the Bonus March in 1932 also did little to endear the Army to the public. Between these public affairs difficulties and the general climate of US public hostility to propaganda in the interwar period, it is perhaps apt to accept the description of Barnet Oldfield, a public relations officer who gained wide experience in World War II, as characteristic of the interwar US Army's approach to public affairs:

> Prior to 1939, the Army had been content to carry on quietly at its posts, a rather clannish society to which few people paid attention. Seldom did top commanders make utterances which would muster headlines, nor did any want

> to do so. In almost any emergency the policy was to play dead or dumb, or both. In all those easy, delicious years, the intelligence officer—by professions and nature the most secretive and non-communicative man on the staff—was considered the logical man to meet the press, if and when they came calling, God forbid! The main function of the intelligence officer in those days was to draft the enemy situation when maneuvers and exercises were being held, and he frequently treated press queries as the field manual said he should answer enemy interrogation—by giving only name, rank and serial number.[89]

With the approach of World War II, it became apparent to many senior officers in the Army that a much more proactive approach to public affairs was necessary to make sure the Army was ready for the massive challenges it would face.

Notes

1. Frederick Palmer, *With My Own Eyes: A Personal Story of Battle Years* (Indianapolis, IN: Bobbs-Merrill, 1933), 339.

2. An Act for Establishing Rules and Articles for the Government of the Armies of the United States, Article 57. http://freepages.military.rootsweb.ancestry.com/~pa91/cfawar.html (accessed 5 May 2009).

3. Jeffrey A. Smith, *War and Press Freedom: The Problem of Prerogative Power* (New York, NY: Oxford University Press, 1999), 92.

4. Phillip S. Paludan, "'The Better Angels of Our Nature': Lincoln, Propaganda, and Public Opinion in the North during the Civil War," in *On the Road to Total War: The American Civil War and the German Wars of Unification, 1861–1871*, ed. Stig Förster and Jörg Nagler (Cambridge, MA: Cambridge University Press, 1997), 357–376.

5. Executive Order, 7 August 1861. John T. Woolley and Gerhard Peters, *The American Presidency Project* [online] (Santa Barbara, CA: University of California (hosted), Gerhard Peters (database)). http://www.presidency.ucsb.edu/ws/?pid=70012 (accessed 5 May 2009).

6. Executive Order, 25 February 1862. Woolley and Peters, *The American Presidency Project.*

7. Benjamin P. Thomas and Harold M. Hyman, *Stanton: The Life and Times of Lincoln's Secretary of War* (New York, NY: Alfred A. Knopf, 1962), 153–155.

8. John Glen, "Journalistic Impedimenta: William Tecumseh Sherman and Free Expression," in *The Civil War and the Press*, ed. David Sachsman, S. Kittrell Rushing, and Debra Reddin van Tuyll (New Brunswick, NJ: Transaction Publishers, 2000), 407–408.

9. James M. McPherson, *Battle Cry of Freedom: The Civil War Era* (New York, NY: Oxford University Press, 1988), 334.

10. John F. Marszalek, *Sherman's Other War: The General and the Civil War Press* (Kent, OH: Kent State University Press, 1999), 37–38.

11. Ibid., 131–132.

12. Ibid., 138.

13. Ibid., 140–141.

14. Ibid., 141.

15. Ibid., 143.

16. Ibid., 144–152.

17. Ibid., 156–159.

18. Indeed, early studies of the Spanish-American War have stressed the degree to which the "yellow journalism" of the day pushed the country into war. For an example of this, see Marcus M. Wilkerson, *Public Opinion and the Spanish-American War: A Study in War Propaganda* (New York, NY: Russell & Russell, 1932). For a more balanced perspective on the role of the press in the war, see Charles H. Brown, *The Correspondents War: Journalists in the Spanish-American War* (New York, NY: Charles Scribner's Sons, 1967).

19. On 2 March 1899, Congress authorized the establishment of a force of 65,000 Regulars supported by 35,000 Volunteers with enlistment terms of 2 years and 4 months. Russell F. Weigley, *History of the United States Army*, enlarged ed. (Bloomington, IN: Indiana University Press, 1984), 290, 308; and Graham A. Cosmas, *An Army for Empire: The United States Army in the Spanish-American War* (Columbia, MO: University of Missouri Press, 1971), 308.

20. Weigley, *History of the United States Army*, 309.

21. Cosmas, *An Army for Empire*, 251–256.

22. Ibid., 256–257.

23. According to Weigley, this letter was initiated by Theodore Roosevelt and Leonard Wood. Weigley, *History of the United States Army*, 310.

24. Cosmas, *An Army for Empire*, 258–263. Given the state of medical knowledge regarding tropical diseases, this was perhaps a less dire number than was perceived at the time.

25. According to Graham Cosmas, Miles "was committed to the theory that the Commanding General owed obedience only to the President and that in war he should have exclusive control over planning and conduct of operations." Cosmas, *An Army for Empire*, 62.

26. Ibid., 285.

27. Ibid., 282–283.

28. Quoted in Cosmas, *An Army for Empire*, 289.

29. Weigley, *History of the United States Army*, 314–320.

30. The Filipino nationalists had maintained a cordon around Manila since the end of fighting with the Spanish in August 1898.

31. Brian McAllister Linn, *The Philippine War 1899–1902* (Lawrence, KS: University Press of Kansas, 2000), 53–55. Indeed, General Otis initially believed the fighting had *not* been ordered by Aguinaldo, but was initiated by some of his more rash subordinates. In his first cable to Washington after news of the attack, Otis wrote, "[I] believe that insurgent army attacked contrary [to the] wishes of their government." He repeated this belief in another dispatch on 7 February. *Correspondence Relating to the War with Spain, Vol. 2* (Washington, DC: Center of Military History, 1993), 894–896.

32. US Army troop strength averaged 40,000, with a peak strength of 70,000 men in December 1900. Linn, *The Philippine War*, 325.

33. Major General Otis replaced Major General Merritt as commander, Department of the Pacific, and military governor of the Philippines on 30 August 1898. Stephen D. Coats, *Gathering at the Golden Gate: Mobilizing for War in the Philippines* (Fort Leavenworth, KS: Combat Studies Institute Press, 2006), 259.

34. Ibid., 69–70.

35. Linn, *The Philippine War*, 134.

36. Ibid.

37. "Confidence Expressed in Otis," *Boston Daily Globe*, 6 February 1899.

38. *Correspondence Relating to the War with Spain, Vol. 2*, 902.

39. Though the US Senate ratified the treaty on 6 February, 2 days after the outbreak of fighting with the Filipino nationalists, it was not signed by the Spanish regent until 19 March. Nonetheless, the return of the US Volunteer units was already becoming an issue. On 3 March, Secretary of War Alger cabled Otis informing him that "As rapidly as possible volunteers should be returned upon ships that bring regulars to you." *Correspondence Relating to the War with Spain, Vol. 2*, 922.

40. "Climax At Hand In The Philippines," *New York Times*, 18 March 1899.

41. "Situation at Manila," *Los Angeles Times*, 18 March 1899.

42. "Otis is a Fussy Old Man," *Boston Daily Globe*, 17 July 1899.

43. For instance, from 18 July 1899, "Newspapermen in Revolt," *Boston Daily Globe*; "Otis' Methods May Undo Him," *Chicago Daily Tribune*; "Otis' Reports Denounced," *New York Times*. The letter was signed by John T. McCutcheon and Harry Armstrong of the *Chicago Record*; Oscar K. Davis and P.G. McDonnell of the *New York Sun*; Robert M. Collins, John P. Dunning, and L. Jones of the Associated Press; John F. Bass and Will Dinwiddie of the *New York Herald*; E.D. Skeene of the Scripps-McRae Association; and Richard Little of the *Chicago Tribune*.

44. Otis to Adjutant-General Corbin, 20 July 1899, *Correspondence Relating to the War with Spain, Vol. 2*, 1036.

45. Linn, *The Philippine War*, 135.

46. Ibid., 206–208.

47. Ibid., 187. Aguinaldo decreed the guerrilla war strategy on 13 November 1899.

48. In April 1901, a number of trials that revolved around the embezzlement of commissary stores began to hit the US papers.

49. "General Smith Admits Charge," *Chicago Daily Tribune*, 26 April 1902, 2.

50. Linn, *The Philippine War*, 321.

51. This incident and the antecedents of Miles' breech with Roosevelt and Root is discussed in Robert Wooster, *Nelson A. Miles and the Twilight of a Frontier Army* (Lincoln, NE: University of Nebraska Press, 1993), 238–241.

52. Edmund Morris, *Theodore Rex* (New York, NY: Random House, 2001), 97.

53. "Gen. Miles Against General Staff Plan," *New York Times*, 21 March 1902, 3; and Wooster, *Nelson A. Miles and the Twilight of a Frontier Army*, 242–243.

54. Morris, *Theodore Rex*, 98.

55. Ibid., 99–104, 127–129; and Linn, *The Philippine War*, 219.

56. Wooster, *Nelson A. Miles and the Twilight of a Frontier Army*, 246–248.

57. "Censorship is in Effect on Mexican Border," *Christian Science Monitor*, 15 March 1916; and "Gen. Pershing Leads Men," *New York Times*.

58. "Censorship Plan Prepared by Army," *New York Times*, 28 May 1916.

59. Woodrow Wilson, "An Address to a Joint Session of Congress," in Arthur S. Link, ed., *The Papers of Woodrow Wilson*, Vol. 41 (Princeton, NJ: Princeton University Press, 1983), 522. For background, see David M. Kennedy, *Over Here: The First World War and American Society* (New York, NY: Oxford University Press, 1980), 10–20.

60. Link, ed., *The Papers of Woodrow Wilson*, Vol. 41, 526.

61. Link, ed., *The Papers of Woodrow Wilson*, Vol. 42, 39–41.

62. Ibid., 59.

63. George Creel, *Rebel at Large: Recollections of Fifty Crowded Years* (New York, NY: G.P. Putnam's Sons, 1947), 158–159. The tensions between Lansing and Creel were never fully overcome. This problem transcended personalities, and has arisen anew almost every time an attempt has been made to create an institutionalized public affairs office beyond the purview of the State Department.

64. "Baker to Censor All Troop News," Special to the *New York Times*, 29 June 1917.

65. "Unconstitutional Censorship Ended: Baker Revokes Order Diverting Cables; 'Emergency Passed' Is Stated Reason," Special to the *New York Times*, 6 July 1917.

66. On the history of the Creel Committee, see Stewart Halsey Ross, *Propaganda for War: How the United States Was Conditioned to Fight the Great War of 1914–1918* (Jefferson, NC: McFarland & Company, 1996); and Stephen Vaughn, *Holding Fast the Inner Lines: Democracy, Nationalism, and the Committee on Public Information* (Chapel Hill, NC: University of North Carolina Press, 1980).

67. John J. Pershing, *My Experiences in the First World War* (New York, NY: Da Capo Press, 1995), 89.

68. Ibid.

69. As was the practice at that time, censorship fell under the purview of Pershing's G-2, LTC Dennis E. Nolan. Frank E. Vandiver, *Black Jack: The Life and Times of John J. Pershing, Vol. II* (College Station, TX: Texas A&M University Press, 1977), 738.

70. For a personal account of his experiences, see Palmer, *With My Own Eyes*.

71. Michael S. Sweeney, *The Military and the Press: An Uneasy Truce* (Evanston, IL: Northwestern University Press, 2006), 54.

72. William M. Hammond, "The News Media and the Military," in *Encyclopedia of the American Military, Vol. III*, ed. John E. Jessup (New York, NY: Charles Scribner's Sons, 1994), 2095.

73. Palmer, *With My Own Eyes*, 340.

74. Paul L. Murphy, "Espionage and Sedition Acts of World War I," in *The Oxford Companion to American Military History*, ed. John W. Chambers II (Oxford, NY: Oxford University Press, 1999), 251–252.

75. Sweeney, *The Military and the Press*, 51.

76. Ibid., 201.

77. Philip M. Taylor, *Munitions of the Mind: A History of Propaganda from the Ancient World to the Present Day*, 3d ed. (Manchester, UK: Manchester University Press, 2003), 183.

78. Raymond Fielding, *The American Newsreel 1911–1967* (Norman, OK: University of Oklahoma Press, 1972), 115–126.

79. David Traxel, *Crusader Nation: The United States in Peace and the Great War, 1898–1920* (New York, NY: Alfred A. Knopf, 2006), 286–287.

80. David Culbert, "Film, War and the Military in," in *Oxford Companion to American Military History*, ed. Chambers, 265.

81. Link, ed., *The Papers of Woodrow Wilson*, Vol. 43, 3.

82. Vaughn, *Holding Fast the Inner Lines*, 197.

83. Ross, *Propaganda for War*, 230.

84. Ibid., 231–240, 317.

85. Ibid., 244–248.

86. Traxel, *Crusader Nation*, 354.

87. Given the large extent to which American financiers continued to participate, even if unofficially, in European postwar reconstruction, one should be cautious about referring to this inward turn as isolationism.

88. Weigley, *History of the United States Army*, 396.

89. Barney Oldfield, *Never a Shot in Anger* (New York, NY: Duell, Sloan and Pearce, 1956), 5.

Chapter 3
World War II

In my past dealings with members of the press and the radio I have scrupulously avoided what might be called propaganda proposals and have endeavored, through a frank presentation of the situation, so far as permissible, merely to give them the facts, leaving the conclusions to their own judgment. The War Department will always be embarrassed by the insatiable demand of our people for "hot" news, and with related perils involved in releasing certain information. The situation is inevitable and the safeguard I turn to is to build up a general understanding of the problems by you gentlemen [American Society of Newspaper Editors] who present carefully considered views in your editorial columns.

George C. Marshall Jr.[1]

After Nazi Germany's invasion of Poland on 1 September 1939, it was apparent to many in the US Government that prudent steps were needed should the United States again be drawn into the wider world war.[2] Given the strong currents of isolationism in American public opinion, the Roosevelt administration faced a daunting task of communicating the rationale for American preparedness and bracing the American populace for potential involvement in the world-spanning conflict.[3] President Franklin D. Roosevelt's own approach to leadership further complicated matters. He had a penchant for fostering bureaucratic rivalries to preserve presidential prerogatives.[4] In keeping with this practice, in the 2 years before American entry into the war, Roosevelt authorized the creation of a number of different agencies tasked with information activities. These included the creation of an Office of Government Reports (established September 1939), a Division of Information of the Office of Emergency Management (established March 1941), an Office of Coordinator of Information (established July 1941), a subsidiary Foreign Information Service (established August 1941), and an Office of Facts and Figures (established October 1941).[5] The growing need to counter German propaganda aimed at Latin America in the summer of 1940 occasioned the appointment of Nelson Rockefeller as the Coordinator of Inter-American Affairs in August 1940. Rockefeller's brief gave him wide latitude in the direction of US public affairs efforts with the region.[6] This confusing profusion of agencies remained in place until the exigencies of wartime prodded Roosevelt to consider a more centralized public information effort.[7]

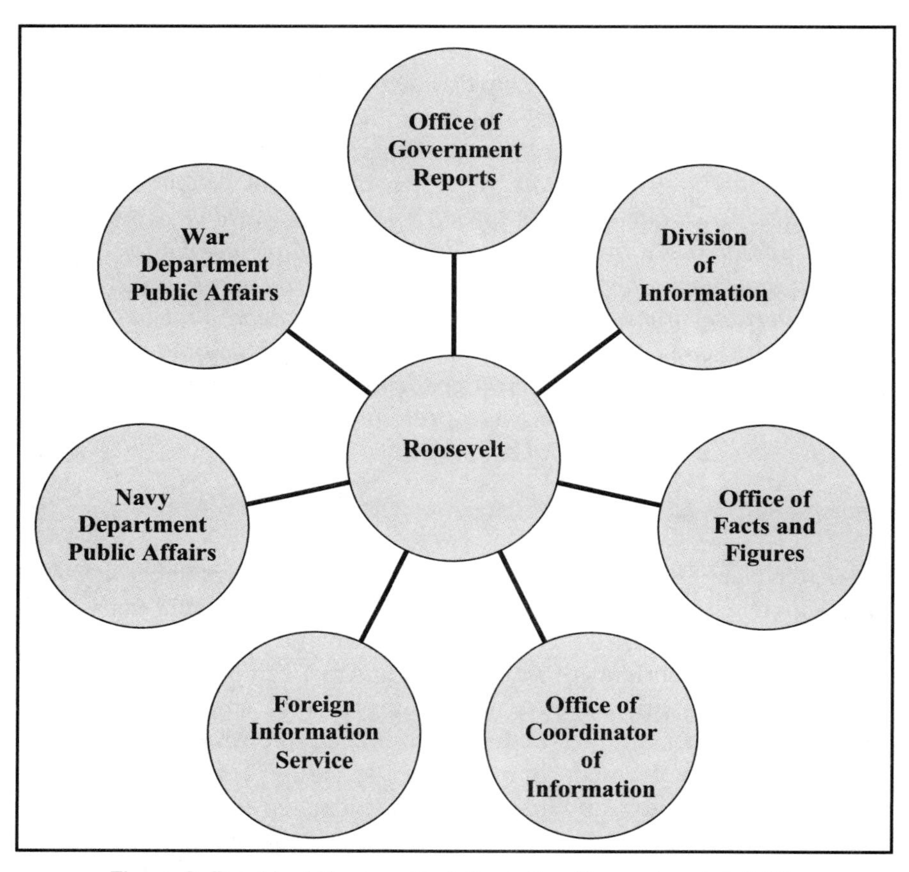

Figure 2. President Roosevelt's Information Soup, circa fall 1941.

The War Department's rudimentary public affairs remained an adjunct of the G-2 Division, where it had been located since 1916. Then, in July 1940, a Press Relation Bureau was established under the Deputy Chief of Staff for Operations. Under this arrangement, the G-2 Division retained responsibility for units located outside of Washington, and the Press Relations Bureau controlled War Department press releases in Washington. Following the fight for the passage of the Selective Service Act, Secretary of War Henry L. Stimson and Chief of Staff George C. Marshall decided that press relations should be more closely coordinated. Stimson decided that a Bureau of Public Relations should be established directly under the Secretary of War and began holding weekly press briefings. Major General Robert C. Richardson Jr. served as the first head of the new Bureau of

Public Relations until August 1941 when he returned to field command. Brigadier General Alexander Surles, who remained in the post throughout the war, replaced Richardson.[8] According to journalist James Reston, by the fall of 1941 the War Department Bureau of Public Relations was staffed with 259 employees, of whom 52 were officers. Both the Navy and War Department's public relations bureaus were working on official budgets of $75,000, exclusive of the salaries. Reston suggested, however, that the amounts employed for public relations work were far higher.[9]

As a corollary to the expansion of official information outlets, the administration also began to review its position regarding censorship. In April 1940, a joint board convened with representatives from the Federal Bureau of Investigation (FBI) and the War, Navy, and State Departments to discuss possible legislation. In early June, the board forwarded to President Roosevelt a "Basic Plan for Public Relations Administration." This plan requested $50 million in appropriations to establish a complete censorship of all mass media outlets. Roosevelt had no interest in pressing for full media censorship, which would have stoked undue political fires. Instead, Roosevelt ordered the War and Navy Departments to compose a plan that limited itself to censorship on international communications.[10] While this plan was being developed, senior officials in both the War and Navy Departments moved to curtail information about the military in public circulation.[11] As per Roosevelt's instructions, a new joint plan for international censorship was submitted and approved by the President on 4 June 1941. This new plan authorized the joint Army-Navy board to prepare legislation that would establish a censorship office for international communications. It also authorized training of Army and Navy personnel for censorship duties in advance of the legislation's passage. Both branches moved quickly to initiate censor training programs. As in World War I, the Army's Censorship Branch remained attached to the G-2 Division of the General Staff. Major W. Preston Corderman served as first head of the Censorship Branch.[12]

Though both the Army and Navy supported leaving censorship in the hands of military authorities, Attorney General Francis Biddle and Secretary of the Treasury Henry Morgenthau Jr. pressed the President to place any office of censorship in civilian hands.[13] On 18 December 1941, Congress approved the first War Powers Act. Among other things, it authorized the President to establish an Office of Censorship. The next day, President Roosevelt appointed Byron Price to direct the new office.[14] Before the war, Price had worked in the wire services and as an executive editor of the Associated Press. Price made a deliberate decision not to follow the pattern of George Creel and the Committee on Public Information. This meant that Price eschewed any responsibilities for shaping Government publicity for fear it would be construed as propaganda. Price's selection also represented a victory for those who felt that the direction of censorship should be retained under civilian control.[15]

The new Office of Censorship released its guidelines for radio and print censorship on 15 January 1942. Journalists and editors were called on to refrain from disclosing information about specific unit and ship names, locations, and movements. A similar policy applied to the movement of the President of the United States and senior military and diplomatic personnel. Weather forecasts and detailed maps and pictures of military installations were also restricted.[16] Despite the fact that the Justice Department had handed down a ruling giving Price wide discretionary powers to control the nation's radio stations and broadcast programs, he opted to follow a policy of voluntary cooperation on the part of the nation's press. This policy was applied to both domestic broadcasts and printed media.[17]

Censorship in war zones remained a military responsibility. This was verified in a joint memorandum issued in late May by the Departments of the Army, Navy, and State along with the Office of Censorship. Under the system put in place for all Navy ships and war zones, journalists had to be accredited. To become accredited, a journalist had to agree to submit all stories to an Army or Navy censor before its release. Theater commanders were responsible for establishing what would be censored in the respective combat zones.[18] Although the War and Navy Departments in Washington established broad guidelines, the theater commanders retained considerable discretion in establishing their public affairs policies. To enter a theater, journalists had to submit to an official examination of their background, discuss their opinions about the war, and prove their reliability to officials in the War and Navy Departments. Once they were accredited, the journalists were subject to military censorship in theater, but in return, the military would feed, transport, and billet them. Combat journalists were also issued officers' uniforms devoid of insignia with green armbands that identified them as photographers or reporters.[19] In the 1930s, United States press and radio organizations had approximately 300 staff correspondents stationed outside the country. During World War II, 1,800 correspondents were accredited to the US Army, and another 800 to the US Navy.[20] In addition, some correspondents operated abroad without accreditation. Though there were individual instances in which these correspondents pressed the boundaries of the censors, in most cases self-censorship by the journalists proved quite effective. This was primarily the case for print and radio journalists. Photographic and newsreel material, on the other hand, were heavily censored throughout the war.[21]

The profusion of government agencies involved in the promulgation of official information and Price's somewhat informal approach to the supervision of censorship contributed to a general unease in both the Roosevelt administration and the press, with the state of affairs, as they existed in the early months of the US entry into World War II. In October 1941, Reston charged that the United States was operating a "vast uncoordinated defense information service which will cost more than $10,000,000 in the current fiscal year." This effort was "run by nobody in particular and by several hundred government officials in general."[22]

Complementing the censorship effort and aiming at coordinating the sprawling range of information agencies, Roosevelt ultimately was convinced to create an Office of War Information (OWI) to consolidate control over the US Government's wartime information programs. OWI was established by Executive order on 13 June 1942. It assumed the duties of the Office of Facts and Figures, the Office of Government Reports, and those duties of the Coordinator of Information and its Foreign Information Service related to gathering public information and its dissemination abroad. President Roosevelt chose Elmer Davis, a civilian, to serve as director. Davis hailed from Indiana, had attended Oxford as a Rhodes Scholar, and later worked for the *New York Times* and as a freelance writer before going on to a successful career as a radio commentator in the late 1930s.[23] Davis faced a difficult task in trying to bring together the activities of the agencies, which had been consolidated into the OWI. This task was made all the more difficult because Davis focused most of his energies on OWI's domestic activities, leaving the overseas component of OWI to Robert Sherwood's direction.

Army Hour

At 1530 eastern time on Sunday, 5 April 1942, the War Department launched one of the most successful public affairs operations in the history of World War II. Secretary of War Stimson touched off the campaign, stating: "Our soldiers in the jungles of Bataan, our airmen in the skies over Burma and the East Indies, have already written immortal pages in the history of the American army. With their spirit inspiring our people, we can and will win the war."[24] During the remainder of the program, Lieutenant General Lesley McNair, chief of Army Ground Forces, and the commanders of the 1st, 2d, 3d, and 4th US Armies sought to assure their listeners that the US military was busy building a "reservoir of trained units ready instantly to go overseas and fight wherever and whenever needed."[25] This broadcast was the first installment of *Army Hour*.

Army Hour was broadcast to the nation over the radio waves by NBC affiliates across the United States every Sunday afternoon. The radio program was developed by the Radio Division of the War Department Bureau of Public Relations. The Radio Division was headed by Edward W. Kirby, a former director of public relations of the National Association of Broadcasters, who was appointed to the rank of lieutenant colonel. Lieutenant Colonel Kirby's division initially consisted of only seven men, though all had previous experience in broadcasting.[26] Major General Alexander D. Surles, chief of the Bureau News Section Colonel R. Ernest Dupuy, and Kirby laid down general guidelines for the program. Wyllis Cooper, a civilian consultant who had previously worked as a radio script-writer, served as writer and director for the series. Cooper had served in the Army along the Mexican border, was gassed at the Argonne in World War I, and remained in the Army as an intelligence officer after the war. He retired from the National Guard in 1933 as a captain.

The program, designed by the Bureau of Public Relations, quickly became a successful Sunday staple. Every week millions of Americans heard the program. After 6 months of broadcasts, *The Washington Post*'s Ernest L. Schier wrote that *Army Hour* was helping "to bring America closer to our allies and to understand what a gigantic job it is to fight all over the world." By weaving together stories from theaters around the globe in addition to the home front, the program seemed to blend home front and fighting front into one.[27] By the end of the first year, in addition to a number of famous guests, several three- and four-star generals, 25 major generals, and 58 colonels had participated in the broadcasts.[28]

After the early reverses suffered by American arms, the Army and the nation at large were eager for positive news stories. The Doolittle raid on Tokyo provided one of the first opportunities to get a positive story out. On 24 May 1942, the country was treated to a radio spot with General James H. Doolittle on the *Army Hour* program. During the interview, which was conducted from an undisclosed location on the West Coast, Doolittle promised that additional Japanese and German targets would be bombed. Doolittle also reassured his audiences that the raid had gone well in large part due to the excellence of the bombers he and his men flew in. According to the Doolittle radio report, "the Japanese pursuit ships that came after us never had a chance."[29] Following up on the positive story in the Pacific, the next week *Army Hour* featured Chinese Nationalist Generalissimo Chiang Kai-shek. In the broadcast, which was translated by his English-educated wife, the Chinese leader promised that China would fulfill its obligations and that the country's "faith in America is unshaken." Chiang also told listeners that he thought the principles of the Atlantic Charter were more than simply "empty diplomatic phrases," but rather principles of freedom, justice, and equality that should be applied to all the people of the world.[30] One of Chiang's lieutenants, Major General Chu Shih-ming, encouraged Americans to believe that numerous airfields in China were available for the deployment of American bombers and aircrews to launch additional raids on Japan.[31]

Subsequent *Army Hour* programs covered a wide-range of subjects. Each week's program came to feature an update on the war presented by Colonel Dupuy. In addition, the technique of using "on-the-spot" broadcasts from around the world was consistently employed. This feature seems to have contributed greatly to the programs' popularity. Vignettes were presented on various aspects of Army training. One week, glider training was featured; another, the life of a West Point cadet.[32] Regular broadcasts from America's allies were also included to give the public a better sense of how war production was being utilized abroad. One broadcast followed a bomber from its construction in Long Beach, California, to a mission over occupied Europe flown out of an airbase in England.[33] A story on the ruggedness of American tanks being used by the British in North Africa was typical of this trend.[34] *Army Hour* also reflected the emotionally charged atmosphere of a nation at war.[35] The headline in *The Washington Post* on 7 December 1942 read, "Marines Find Killing Japs on Fly Great Sport." In the broadcast on the previous day, Colonel Leonard Rodieck described the practice by which the Marines dealt with Japanese snipers in the South Pacific. If a sniper was found in a coconut tree, a tank would be called up. It would then bump the tree, not hard enough to uproot the tree, but rather to dislodge the sniper so that he would be hurtled "through the air in much the same manner as a clay pigeon." A Colonel Thomas, serving as chief of staff to General Alexander A. Vandergrift, reported over the airwaves that "shooting a Jap on the fly is the most popular sport on Guadalcanal at the moment."[36] However, while such a report would be sure to offend contemporary sensibilities, it was not unusual given the racially charged context of the war in the Pacific.[37]

Army Hour retained its position as one of the most successful and sustained programs for the remainder of the war. The program "reported on the Army in its darkest moments—at the surrender of Corregidor, during the bad days at Anzio, at the Ardennes breakthrough. But they were also there to hail the Army in its brightest successes."[38] The Army and, later, the Department of Defense would continue to try to emulate its success with programs like *Time for Defense* and *Battle Report, Washington*.[39]

Military-Media Relations in North Africa, the Mediterranean, and Europe

In the fall of 1942, General Dwight D. Eisenhower, the designated US commander of the landings in North Africa, found a number of challenges in his relations with the press. He was initially intent on retaining strict operational security, lest "the effort to keep our intentions secret and to give credence to our cover and deception plans will be hopeless."[40] At the same time, Eisenhower instructed Army censors not to block anything that was critical of him personally.[41] In mid-September Eisenhower repeatedly expressed his concern to General George Marshall that the press was using "detective methods" to ferret out his operational plans. His expressed his frustration, perhaps tongue-in-cheek, in a letter on 19 September, in which he wrote: "In a high-pressure situation, such as now prevails here, the knowledge that all our efforts may be defeated by some damnable and inexcusable act on the part of the press is peculiarly upsetting. It would be a great pleasure to hang the offender!"[42] Eisenhower's sensitivity is understandable, especially given the fact that he faced the complications of both coalition warfare and planning for a series of amphibious landings that needed to occur nearly simultaneously over a vast geographic swath of territory.

The planning for Operation TORCH was fraught with both military and political difficulties, not least of which was ascertaining how the French authorities in North Africa would react to any allied landings.[43] The North African campaign would not only provide units of the US Army with a baptism by fire, they also introduced Eisenhower to the complications of diplomacy and public relations that befall theater commanders. These issues came sharply into focus over the attempts to make a deal with Admiral Francois Darlan, commander in chief of the Vichy French military forces, in the wake of the TORCH landings. The Americans and British had initially pursued a deal with General Henri Giraud, a French patriot who had fought the Germans in World War I and World War II before being captured, escaping, and going into hiding in the south of France.

The basis of a deal with Giraud had been discussed by Robert Murphy, then US Counsel-General to French North Africa, and French officers sympathetic to the allied cause, including Major General Charles Mast, then serving as Chief of Staff to the French 19th Corps in Algiers. As a condition of cooperation, Mast insisted on receiving information on the allied invasion plan. Major General Mark Clark, then serving as Eisenhower's American Deputy Commander, and a small group from Eisenhower's Allied Forces Headquarters (AFHQ) staff carried out this delicate task by means of a covert operation. Clark and his team flew to Gibraltar, and were then conveyed by a British submarine from Gibraltar to the coast just outside of Algiers. Clark's team went ashore in the morning hours of 22 October 1942. During this meeting, Mast encouraged Clark to believe that if General Giraud could be spirited from his hiding place in the south of France to North Africa, he would rally French forces and prevent bloodshed when the American landings took place.[44] Giraud was indeed spirited away from France by submarine, but his insistence on assuming command of the Allied forces was unacceptable to either Roosevelt or Winston Churchill. In addition, Giraud's influence over the French forces in North Africa was less than the allies had hoped. Nonetheless, the War Department subsequently used Clark's secret trip for its public affairs value as a tale of high adventure.

When the TORCH landings did take place on 8 November, Robert Murphy, Roosevelt's political representative in North Africa, tried to secure cooperation from General Alphonse Juin, the commander in chief of French forces in North Africa.[45] By chance, Juin's superior, Admiral Darlan, happened to be in North Africa visiting his son. Darlan initially proved unwilling to countenance cooperation. Eager to prevent fighting with the French and under pressure from Churchill to secure the French Navy, Eisenhower dispatched Clark to Algiers to negotiate with Darlan. It took another 2 days of allied diplomatic pressure before Darlan approved a cease-fire in return for allied recognition as the de facto French military governor of North Africa. When news of the cease-fire was broadcast, the Germans reacted by promptly occupying Vichy.[46]

In the interim, General Marshall cabled Eisenhower and informed him that while the operation unfolded, he wanted the press to be kept busy with positive news stories. He specifically suggested that the story of Clark's secret rendezvous with the French military authorities outside Algiers would make for good copy.[47] In his reply to Marshall, Eisenhower demonstrated a growing appreciation for information management. He informed Marshall that he had hitherto withheld the information out of concern that prior to securing French North Africa, the story would be exploited by the Axis. This could have been done by implying that the Americans had to undertake a secret mission to meet with "Quislings" within the Vichy French military. By releasing the story after the landings had taken place and any fighting had ended, Eisenhower hoped to portray the mission as "immensely important to us in finding out exactly what was the majority sentiment in North Africa and in preparing the way for effective United States-French collaboration."[48] The story did make quite a splash, appearing in numerous papers. The front page of the *New York Times* carried the headline "A Spy-Thriller Trip by Clark Paved Way for Africa Drive."[49] Despite the positive spin the story generated, neither Marshall nor Eisenhower were completely pleased.[50] According to Marshall, reports of Clark's boat capsizing, which led to both the general's pants and the gold he was carrying being lost at sea, "cheapened" the efforts of the American officers. Worse yet, the discussion of the amount of gold lost threatened to point to American bribery of French officers.[51] Clearly this threatened to undermine the premise that the Americans were establishing contacts with pro-allied elements in the French military.

General Eisenhower soon found himself facing a new information management problem. Once the Axis forces had been expelled from North Africa and the decision made to press on to Sicily, he had to work to defuse speculation about the next stage of Allied military operations. What made this difficult were the extensive preparations underway for the Sicilian campaign across North Africa. Exercises were being held on numerous beaches and equipment and landing craft were being gathered in ports. Eisenhower's concern was that, given the absence of combat activity, reporters would turn to speculation about the future course of Allied operations. Given the intense scrutiny the Allied preparations were under by Axis intelligence officers, General Eisenhower took the unprecedented step of bringing the reporters into his confidence. A month before the launch of Operation HUSKY, Eisenhower briefed his press corps on the invasion, indicating the intent to attack Sicily, spelling out the general locations of British General Sir Bernard Montgomery and Lieutenant General George S. Patton's landing beaches, and indicating that the Allies were heavily bombing the western end of Sicily to mislead the Axis as to the intended location of the landings. According to Eisenhower, "from that moment onward, until after the attack was launched, nothing speculative came out of the theater and no representative of the press attempted to send out anything that could possibly be of any use to the enemy."[52] Though this was an experiment that Eisenhower, given his responsibility for maintaining operational secrecy, did not want to repeat, he was cognizant that it "placed upon every reporter in the theater a feeling of the same responsibility that I and my associates bore."[53]

On 15 January 1944, General Eisenhower arrived in London to assume command of Supreme Headquarters Allied Expeditionary Force (SHAEF) and hasten the planning for the Allied invasion of northwest Europe scheduled for 1 May 1944. It was clear to all involved that it would be impossible, given the magnitude of the buildup, to prevent the Germans from realizing that an invasion of France was impending. Nonetheless, there was clear military need to achieve as much strategic and tactical surprise as possible to maximize the operation's chance of success. Indeed, a strategic deception plan had already been deemed necessary at the Cairo Conference the previous autumn. SHAEF would subsequently develop a plan, known as Operation FORTITUDE, which proved remarkably successful in convincing the Germans that the main Allied attack would take place as the Pas de Calais.[54] Thus, even after word of the Normandy landings, the Nazi high command hesitated to commit reinforcements for a number of days on the supposition that the main assault was still coming. To pull this off, a network of double agents, fictitious radio traffic, and elaborately staged encampments and supply depots were established in southeast England. A nonexistent First United States Army Group (FUSAG), nominally under the command of Patton, was also conjured into existence to make the Germans believe the flamboyant Patton would be leading the first wave of the assault.

These elaborate preparations, in part, prevented these deception and cover operations from being divulged. With hundreds of reporters pouring into the United Kingdom, maintaining a good working relationship with the media entered into a critical phase for the maintenance of operational security. At his first press conference as Supreme Allied Commander on 17 January, Eisenhower impressed on the assembled journalists that once they were accredited to his headquarters, he considered them to be quasi-staff officers. On 28 January, Prime Minister Churchill wrote to General Eisenhower expressing his concern over the maintenance of security for Operation OVERLORD. Churchill informed Eisenhower that he had seen to it that the British and American press (through the American censors who worked in cooperation with the British Ministry of Information) in Britain took care not to "forecast dates, areas of landings, or estimates of possible size of attacks," but was concerned that as journalists became accredited to SHAEF, similar security precautions needed to be followed.[55] Eisenhower replied that he personally would "feel disturbed if I thought that I or my Public Relations Staff were held as anything but friends of the Press." Eisenhower was in no rush to prematurely accredit press individuals before the invasion out of concern that it might stimulate a "Thank you for nothing" sentiment. By the end of May, Sir Cyril Radcliffe, Director General of the British Ministry of Information, and Brigadier General Robert A. McClure, Eisenhower's head of public relations at SHAEF, worked out an agreement to begin quietly accrediting journalists.[56]

Subsequently, many of these accredited journalists would be assigned to accompany specific outfits once the invasion began. With a limited number of spots available on the first assault wave, some journalists were offered the opportunity to parachute in with the airborne elements. Though it was a tough sell, a number of journalists ultimately found their way into France by this route.[57] On 8 May 1944, Eisenhower issued guidance for handling accredited members of the press to all unit commanders in the AEF stating: "As a matter of policy accredited war correspondents should be accorded the greatest possible latitude in the gathering of legitimate war news." For Eisenhower, subject only to the needs for operational security judged by the individual commanders, the journalists were to have access to visit with officers and enlisted men, as well as "see the machinery of war in operation in order to visualize and transmit to the public the conditions under which the men from their countries are waging war."[58] For those journalists who were embedded with paratroopers or units in the initial assault wave, their experience in Normandy began in a most visceral manner. Those who remained behind in England met at Macmillan Hall at the University of London. There they were locked in the Press Room, furnished with background material, and at 0830 read the famous terse official communiqué, "Under the command of General Eisenhower, Allied naval forces, supported by strong air forces, began landing Allied armies this morning on the northern coast of France."[59] Although SHAEF accredited 530 journalists by 7 June 1944, prior to the liberation of Paris there was a limit to the number of press representatives allowed on the continent. After the liberation of Paris, this system was ended.[60] By war's end, SHAEF would accredit 1,338 correspondents from the United States, Britain, France, and some neutral nations.[61]

With the approach of war's end, both the Office of Censorship and the Office of War Information were closed. In the case of the Office of Censorship, Director Byron Price had long advocated that censorship be curtailed as soon as the threat to national security decreased. He advocated the closure of his agency as soon as fighting ended. The Censorship Policy Board endorsed Price's views on 20 November 1943. After the defeat of Germany, Price "canceled all program restrictions of the radio code." He subsequently won President Truman's endorsement to declare the end of censorship on the same day that victory over Japan was announced. The work of the Office of Censorship formally ended on 15 August 1945. Price's plan for a voluntary censorship code was generally perceived to have worked admirably well. The editors of the trade journal *Editor & Publisher* stated, "We have never heard anyone in the newspaper business contradict the statement that Byron Price conducted the Office of Censorship in a competent, careful and wholly patriotic manner."[62]

While there had probably never been any question of the Office of Censorship continuing its activities in peacetime, a reasonable case could be made for the retention of the Office of War Information. Unlike the

Creel Committee, which had focused the great majority of its propaganda output to maintain domestic support for President Wilson's policies, from 1944 on, 90 percent of OWI's budget was geared toward international propaganda activities.[63] At the end of August, President Truman signed another Executive order that transferred the foreign information functions of the Office of War Information and the Office of Inter-American Affairs to an Interim International Information Service established within the Department of State. Under this order, the remaining functions of the Office of War Information were to cease on 15 September 1945. As of 31 December 1945, both the Office of War Information and the Interim International Information Service were abolished.[64] The War Department retained its Public Relations Bureau, the function of which was increasingly geared toward defending Army appropriations during the massive postwar demobilization. This quiet interlude for Army public affairs, discussed in the next chapter, did not last long. The dawning of the Cold War would soon revive the interest of the US Government in pursuing a more aggressive approach to public affairs, public diplomacy, and psychological warfare. In the context of the Cold War, the military found itself called on to fight limited wars very different from World War I and World War II. The difficulties of limited war were manifested in a number of ways, including creating a complicated environment for military-media relations.

Notes

1. Informal and Off-the-Record Statement to the American Society of Newspaper Editors, 13 February 1943, *Marshall Papers* [Microfilm], Reel 24.

2. On the Army's prewar preparations, see Mark Skinner Watson, *Chief of Staff: Prewar Plans and Preparations* [United States Army in World War II, The War Department] (Washington, DC: Center of Military History, 1991), 126–176.

3. For a provocative assessment of this challenge, see Steven Casey, *Cautious Crusade: Franklin D. Roosevelt, American Public Opinion, and the War against Nazi Germany* (Oxford, NY: Oxford University Press, 2001).

4. William Leuchtenburg wrote: "In flat defiance of the cardinal rule of public administration textbooks—that every administrator appear on a chart with a clearly stated assignment—the president not only deliberately disarranged spheres of authority but also appointed men of clashing attitudes and temperaments." William E. Leuchtenburg, "Franklin D. Roosevelt: The First Modern President," in *Leadership in the Modern Presidency*, ed. Fred I. Greenstein (Cambridge, MA: Harvard University Press, 1988), 28.

5. Allan M. Winkler, *The Politics of Propaganda: The Office of War Information, 1942–1945* (New Haven, CT: Yale University Press, 1978), 21–28.

6. Ibid., 25. On Roosevelt's concern regarding Nazi activities in Latin America, see Robert Dallek, *Franklin D. Roosevelt and American Foreign Policy, 1932–1945* (New York, NY: Oxford University Press, 1979), 233–236.

7. Winkler, *The Politics of Propaganda*, 28–29.

8. Forrest C. Pogue, *George C. Marshall: Organizer of Victory* (New York, NY: The Viking Press, 1973), 127–128.

9. James B. Reston, "Defense Picture Given to World in All Media by Federal Agencies," Special to the *New York Times*, 27 October 1941.

10. Richard W. Steele, "Preparing the Public for War: Efforts to Establish a National Propaganda Agency, 1940–41," in *The American Historical Review* 75, No. 6 (October 1970): 1641–42; and Michael S. Sweeney, *Secrets of Victory: The Office of Censorship and the American Press and Radio in World War II* (Chapel Hill, NC: University of North Carolina Press, 2001), 27–28.

11. Sweeney, *Secrets of Victory*, 22–27.

12. Ibid., 28.

13. Ibid., 30.

14. The Office of Censorship was officially established under Executive Order 8985. Sweeney, *Secrets of Victory*, 11, 36.

15. Ibid., 45–46.

16. The summary text of the new code was published in "Censorship Code Issued for Press," Special to the *New York Times*, 15 January 1942.

17. Sweeney, *Secrets of Victory*, 7–10.

18. Ibid., 51.

19. Ibid., 106–107.

20. Robert W. Desmond, *Tides of War: World News Reporting, 1940–1945* (Iowa City, IA: University of Iowa Press, 1984), 448–449.

21. Raymond Fielding, *The American Newsreel, 1911–1967* (Norman, OK: University of Oklahoma Press), 288–295; and George H. Roeder Jr., *The Censored War: American Visual Experience during World War II* (New Haven, CT: Yale University Press, 1993).

22. Reston, "Defense Picture Given to World," Special to the *New York Times*, 27 October 1941.

23. Clayton D. Laurie, *The Propaganda Warriors: America's Crusade against Nazi Germany* (Lawrence, KS: University Press of Kansas, 1996), 110–111; and Winkler, *The Politics of Propaganda*, 31–35.

24. "Victory Keynote Opens 'Army Hour,'" Special to the *New York Times*, 6 April 1942.

25. Ibid. Quote from Lieutenant General McNair. For an anecdotal account, see Edward M. Kirby and Jack W. Harris, *Star-Spangled Radio* (Chicago, IL: Ziff-Davis Publishing, 1948), 25–41.

26. Gerd Horten, *Radio Goes to War: The Cultural Politics of Propaganda during World War II* (Berkeley, CA: University of California Press, 2002), 41.

27. Ernest L. Schier, "'Army Hour' Radio Program Ranks In Importance With Fighting Fronts," *The Washington Post*, 11 October 1942.

28. "The Army Hour: A Year of Sundays," *New York Times*, 4 April 1943.

29. "Doolittle Promises More Raids on Japan," *New York Times*, 25 May 1942; and "Doolittle Sees Raids on 'Other Jap Objectives,'" *Washington Post*, 25 May 1942.

30. "Chiang Ties Morale And Arms," *Christian Science Monitor*, 1 June 1942.

31. "Air Bases in China," *Christian Science Monitor*, 1 June 1942.

32. "Glider Training To Be Subject Of Army Hour," *Chicago Daily Tribune*, 7 June 1942; and "A West Point Plebe Is Guest on Army Hour," *Chicago Daily Tribune*, 26 July 1942.

33. "How U.S. Fights Enemies Shown By Army Hour," *Chicago Daily Tribune*, 19 June 1942.

34. "New U.S. Tank Outhit, Outlast Nazis in Libya," *Washington Post*, 8 June 1942.

35. For a contemporary analog, see "US Sniper shot at Koran in Iraq," BBC News, 18 May 2008. http://news.bbc.co.uk/2/hi/middle_east/7407187.stm (accessed 14 May 2009).

36. "Marines Find Killing Japs on Fly Great Sport," *Washington Post*, 7 December 1942.

37. John W. Dower, *War Without Mercy: Race and Power in the Pacific War* (New York, NY: Pantheon Books, 1986).

38. Kirby and Harris, *Star-Spangled Radio*, 40–41.

39. Steven Casey, *Selling the Korean War: Propaganda, Politics, and Public Opinion in the United States, 1950–1953* (Oxford, NY: Oxford University Press, 2008), 91, 104.

40. Alfred E. Chandler, ed. *The Papers of Dwight David Eisenhower, The War Years, Vol. I* (Baltimore, MD: The Johns Hopkins Press, 1970), 564.

41. Harry C. Butcher, *My Three Years with Eisenhower* (New York, NY: Simon and Schuster, 1946), xiv–xv.

42. Chandler, ed., *The Papers of Dwight David Eisenhower, The War Years, Vol. II*, Doc. 509, 565

43. Carlo D'Este, *Eisenhower: A Soldier's Life* (New York, NY: Henry Holt and Company, 2002), 343–346.

44. Martin Blumenson, *Mark Clark: The Last of the Great World War II Commanders* (New York, NY: Congdon & Weed, 1984), 81–83.

45. Robert Murphy, *Diplomat Among Warriors* (Garden City, NY: Doubleday & Company, 1964), 124–134.

46. D'Este, *Eisenhower*, 354.

47. Pogue, *George C. Marshall: Ordeal and Hope,* 418–419.

48. Chandler, ed., *The Papers of Dwight David Eisenhower, The War Years, Vol. II,* Doc. 603, 686.

49. "A Spy-Thriller Trip by Clark Paved Way for Africa Drive," *New York Times*, 13 November 1942.

50. Though he did not complain to Marshall, Eisenhower had specifically requested in an earlier cable that the means of Clark's transport not be disclosed, because submarine insertions were widely used in covert operations. A War Department press release on 13 November, however, specifically mentioned, "Prior to the beginning of current operations, General Clark, with a few trusted assistants, proceeded secretly in a submarine to enemy-dominated territory." Quoted in *The Washington Post*, 13 November 1942. On Eisenhower's request that the mode of transportation not be mentioned, see Chandler, ed., *The Papers of Dwight David Eisenhower, The War Years, Vol. II*, Doc. 603, 686.

51. Chandler, ed., *The Papers of Dwight David Eisenhower, The War Years, Vol. II*, Doc. 652, 749, No. 1.

52. Dwight D. Eisenhower, *Crusade in Europe* (Garden City, NY: Doubleday & Company, 1948), 169–170.

53. Ibid.

54. Roger Hesketh, *Fortitude: The D-Day Deception Campaign* (London: St. Ermin's Press, 1999).

55. Churchill's letter is referenced, but not reproduced, in Chandler, ed., *The Papers of Dwight David Eisenhower, The War Years, Vol. III*, Doc. 1532, 1709, No. 1. The British Ministry of Information was established on 3 September 1939. Like the US Creel Committee in World War I, it combined the functions of censorship, domestic information, and propaganda in a single organization. Michael Balfour, *Propaganda in War, 1939–1945: Organisations, Policies and Publics in Britain and Germany* (London: Routledge & Kegan Paul, 1979), 53–71.

56. Chandler, ed., *The Papers of Dwight David Eisenhower, The War Years, Vol. III*, Doc. 1532, 1709, No. 2.

57. The first journalist to jump into combat reporting was Jack Thompson of the *Chicago Tribune*. He parachuted in with Lieutenant Colonel Edson Raff's battalion at Youkes-les-Baines in Algeria on 15 November 1942. Barney Oldfield, *Never a Shot in Anger* (New York, NY: Duell, Sloan and Pearce, 1956), 47–58.

58. Chandler, ed., *The Papers of Dwight David Eisenhower, The War Years, Vol. III*, Doc. 1676, 1853.

59. Eisenhower's Communiqués Digital Collection, Communiqué Number 1, 6 June 1944. http://contentdm.lib.byu.edu/cdm4/document.php?CISOROOT=/EisenhowerCommuniques&CISOPTR=652&REC=2&CISOSHOW=0 (accessed 14 May 2009).

60. Forrest C. Pogue, *The Supreme Command* (Washington, DC: Center of Military History, US Army, 1989 reprint), Appendix A: SHAEF and the Press, June 1944–May 1955, 522.

61. Desmond, *Tides of War*, 449. Pogue gives the total number of accredited journalist as 996 "shortly before the war's end," but this may not account for the total number of unique individuals who had entered the theater. Pogue, *The Supreme Command*, 449.

62. Sweeney, *Secrets of Victory*, 209–210.

63. Shawn J. Parry-Giles, *The Rhetorical Presidency, Propaganda and the Cold War, 1945–1955* (Westport, CT: Praeger, 2002), 4.

64. Executive Order 9608, 31 August 1945. John T. Woolley and Gerhard Peters, *The American Presidency Project* [online] (Santa Barbara, CA: University of California (hosted), Gerhard Peters (database)). http://www.presidency.ucsb.edu/ws/?pid=60671 (accessed 5 May 2009).

Chapter 4

Murky Waters: Military-Media Relations and Limited Wars

Among all the contemporary skills with which a soldier these days must concern himself, not the least important is public relations—a phrase almost unknown to the Army and a profession little practiced by it until World War II. That ignorance or negligence may be one reason why at the end of every war the Army was a budgetary stepchild. Chief of Staff might present and even argue their views that appropriations were inadequate—but they did it to their civilian superiors or to congressional committees. The general public, either as an interested audience or as a support, was largely ignored because of a long run tradition, accepted by the Army, that soldiers should be seen and not heard.

Dwight D. Eisenhower[1]

During World War II, Secretary of War Henry Stimson and Chief of Staff of the Army George C. Marshall had kept close watch on the US Army's public relations. When Marshall retired on 26 November 1945, his long-time protégé, Dwight D. Eisenhower, succeeded him as Chief of Staff. For the next 3 years, Eisenhower was largely preoccupied with postwar demobilization, interservice fights over unification, and the debate over the need for universal military training. Eisenhower's experience in North Africa and Northwest Europe gave him considerable experience in the diplomacy of coalition warfare and an appreciation for information management. Like Stimson and Marshall, Eisenhower had become adept in his relations with the press during the war. As Chief of Staff, Eisenhower recognized that the need for an Army public relations capability was perhaps more important than ever. Eisenhower's wartime experience also convinced him that "the expenditure of men and money in wielding the spoken and written word was an important contributing factor in undermining the enemy's will to resist and supporting the fighting morale of our potential Allies in the Occupied Countries." This led him to the conclusion that "without doubt, psychological warfare has proved its right to a place of dignity in our military arsenal."[2] Eisenhower was not alone in his convictions. As the postwar peace shifted to the Cold War confrontation with the Soviet Union, the US Government would pay increasing attention to the improvement of its propaganda and psychological warfare capabilities.

Before all that happened, however, Eisenhower appointed Lieutenant General "Lightnin' Joe" Collins, in charge of the Army's public affairs capabilities. Collins, who had distinguished himself during the war as commander of the 7th Corps from the Normandy landings to the German surrender, was initially ill-disposed to give up field command to serve as the War Department's Chief of Information. Judging by Collins' own accounts, he was never entirely comfortable in his shift from the field to public relations.[3] Collins recognized that "it is the job of the public relations officer to assist the commander in cementing this partnership with the public by providing accurate, full and unbiased information." For Collins the role of the public relations officer also included the indoctrination of recruits and "by interpreting the profession of arms to a nation which is eager to be proud of its Army."[4] One example of Collins' somewhat rocky relations with the press came in the summer of 1946. A rash of pregnancies among nurses had resulted in rather negative comments about the American occupation forces. At Eisenhower's behest, Collins traveled to Germany to inform senior commanders of the need to maintain a high standard of conduct. When Collins addressed the press on the matter from the American occupation headquarters in Frankfurt, he accused the American reporters of "concentrating on scandal instead of important aspects of the Cold War being fought between the Russian and Allied forces of occupation."[5] Though Collins' salvo probably did little to mute the interest in covering such stories, he found that his experience as Chief of Information was an ideal school to prepare a senior officer for the challenges that a Chief of Staff of the Army faced.[6]

As Chief of Information, Collins authorized an independent study of the War Department public affairs practices by Jack H. Lockhart. Lockhart was an assistant to John H. Sorrells, an executive editor for the Scripps-Howard Newspapers. During World War II, Lockhart had served as assistant director of the press section in the Office of Censorship. His report was submitted to Collins and Major General Floyd Parks in early June 1946.[7] Lockhart opened by pointing out that the Army would be a source of news with or without its cooperation. In Lockhart's view, the surest way to guarantee that news would be unfavorable would be to try to choke off the bad news. This would only "increase the news value of the unfavorable and decrease interest in the favorable."[8] In final analysis, Army public relations could only "present the Army as it is, not as it ought to be or would like to be. No more than that can be expected, or achieved."[9] Lockhart realized that a "warts and all" approach was unpalatable to many who desired the public to perceive the Army as perfect or ideal. But for

Lockhart, "this desire for a paper army of perfection only is a sure course to an army rotten in fact."[10]

On the practical side, Lockhart had a number of recommendations for Army public relations. He believed the four words for Army public affairs officers to live by were accessibility, frankness, speed, and authority.[11] He felt that the Secretary of War and the Chief of Staff had to be the foremost public spokespersons of the Army, and that any announcement or information they released should be quickly conveyed to the Bureau of Public Relations. Anyone who served as spokesperson had to be of sufficient authority that his statement was unlikely to be reversed by someone up the chain of command. Lockhart thought that the Director of Information and the Director of the Bureau of Public Relations should participate in all War Department policy planning and decisions. He recommended the establishment of officers in the Bureau of Public Relations who would monitor and be available to respond rapidly to any breaking news developments affecting the Army.[12] Lockhart argued that all Army responses to inquiries must be the whole, unvarnished truth. Not only did he warn against half-truths, he recommended eliminating the practice of buck passing and banishing the phrase "no comment." The latter, he thought, would certainly goad reporters to seek their story elsewhere, probably to the Army's regret.[13] Lockhart also suggested what might be termed a traveling inspector general of public relations who would visit with Public Relations Officers (PROs) in the field and identify commanders who needed further education on the importance of public relations.[14] Finally, he encouraged PROs to read media trade journals, such as *Editor and Publisher*, and high-level Bureau of Public Affairs officers to maintain friendly relations with managing editors of the major papers, the National Press Club, and the Gridiron Club. For Lockhart, public relations needed to be a "command mission throughout the Army" because "command indifference, neglect, ignorance, abuse or opposition to public relations can make futile the work of the most able and hard-working PRO or public relations group, and quickly tear down more good will than subordinates can ever build up."[15]

In 1946, the Army launched two initiatives to improve its public affairs capability. The Army Information School was established at Carlisle Barracks, Pennsylvania.[16] Brigadier General Williston Palmer served as the first commandant of the school. Two tracks of instruction were established. The first focused on public affairs and the other on information and education. The latter aimed at training officers who would provide information to soldiers currently serving regarding the purpose of their service. Both began with 8-week courses that enrolled 100 officers each.

The curriculum of both included refresher material on American history and discussion of the functions of the War Department and its components. In addition, considerable time was devoted to practical aspects of the job, to include giving public speeches and the preparation of press releases, visual aids, and publications. The Director of Information established and published a new journal, the *Army Information Digest*. The journal aimed to provide information to commanding officers and provide useful background to public affairs officers to aid in the discharge of their duties.[17]

After the establishment of the position of Secretary of Defense under the National Security Act of 1947, the first incumbent, James Forrestal, advocated the centralization of many defense functions, both to prevent needless duplication and to curb interservice disputes over roles and missions. In this vein, he established the nonstatutory Office of Public Information (OPI) on 17 March 1949. It was Forrestal's expectation that OPI would serve as the sole public relations outlet for the military services in Washington, DC.[18] OPI was charged with responsibilities for the press (print, radio, newsreels, and television), the accreditation of correspondents, analysis of public information pertaining to the military, and review of materials (such as manuscripts) for publication. OPI soon came under determined resistance from the individual Services, which resisted the amalgamation of their individual public affairs capabilities. In March 1950, the Service secretaries requested that Secretary of Defense Louis Johnson restore public affairs capabilities to the individual Services.[19] Though one would suspect that the outbreak of the Korean War would have necessitated a strengthening of OPI—this did not occur. Paradoxically, the impact of the Korean War allowed the individual Services to more strongly assert their individual public affairs capabilities, while the White House attempted to assert closer control over the release of information pertaining to foreign and military policy.[20]

The White House's interest in closer control over information management largely resulted from the frustration caused by General Douglas MacArthur's frequent, unwarranted broadsides on matters of policy beyond his own purview, especially his pronouncements on US strategic intentions regarding Formosa and mainland China. As in all things, MacArthur's relations with the press over the course of his career provided a diverse array of experiences. Before the US entry into World War I, MacArthur had served as the War Department's one-man Bureau of Information. In 1916, MacArthur held the firm conviction that the Government and the military needed a policy of strict censorship should the United States enter the war. He advocated strict censorship because he thought the press might betray military secrets and demoralize the American public as a result of bias or inaccurate reporting.[21] As Chief of Staff of the Army (1930–35), MacArthur had been directly responsible for one of the worst publicity disasters the interwar Army experienced when he personally took charge of the operation that drove the Bonus Army from the streets of Washington, DC. Though MacArthur was keen to cultivate a small circle of admiring press correspondents, his handling of the press during World War II alienated many.

During the defense of the Philippines, an initial promise to hold daily press conferences broke down as the fighting went poorly. In lieu of the press conferences, MacArthur relied on tight censorship of reporters. With this tight censorship in effect, the reporters in the Philippines were forced to rely on communiqués issued by MacArthur's aide, Colonel Legrande Diller. According to Diller's later assertion, MacArthur himself wrote or closely edited most of these communiqués. Not surprisingly, MacArthur received the lion's share of attention in the communiqués. Between 8 December 1941 and 11 March 1942, 109 of the 142 issued communiqués only referenced one individual—MacArthur. The names of combat units were generally omitted, and the men doing the fighting were referred to as "MacArthur's army" or "MacArthur's men." D. Clayton James has pointed out that these releases were essentially propaganda, though it was uncertain what MacArthur hoped to accomplish by them. In any event, they "did not lessen the misunderstandings between him and his troops or officials in Washington."[22] MacArthur's constant interest in publicity contributed to poor relations with the US Navy. Though this was only one element in the larger struggle between Admiral Chester W. Nimitz and MacArthur to establish the strategic priorities for the war in the Pacific, MacArthur's desire to manage news to his own benefit clashed with the Navy's desire to maintain the tightest possible operational secrecy regarding what types and how many ships were involved in any given action.[23] In his handling of the communiqués for the campaign in New Guinea, James argues that MacArthur was trying to channel the primary Allied war effort into his theater in the Pacific.[24]

When the war in the Pacific ended, journalists from America and other Allied nations were eager to see wartime censorship and with it their dependence on communiqués cease. Stories about the Japanese surrender and subsequent occupation promised ample opportunities for "scoops." In Europe, General Eisenhower had received the German surrender in the middle of the night with no fanfare. When General MacArthur had been designated Supreme Commander Allied Powers (SCAP) by President

Harry S. Truman on 15 August, he quickly decided to carry out an elaborate and public formal surrender.[25] Over the next 2 weeks, the details of the surrender ceremony were worked out. MacArthur, in an effort to salve the Navy's wounded pride when Nimitz was not made SCAP and in a nod to Truman, chose the battleship *USS Missouri* for the site of the famous ceremony on 2 September 1945 that marked the formal end of World War II.

Much to the chagrin of many Western journalists, however, some wartime practices lingered into the occupation. To add insult to injury, despite formal prohibitions from the occupation authorities, some Japanese radios continued to broadcast in English overseas after the surrender. In several instances, Japanese radios broadcasted news of decisions from the American occupation authorities many hours before MacArthur's staff relayed the news to the Western press.[26] MacArthur also made statements—having been duly alerted to the political implications of getting the boys home soon—that conflicted with the War Department and the Truman administration's policy on demobilization.[27] When Emperor Hirohito made his unprecedented call on MacArthur at the US Embassy, bayonet-wielding troops prevented the entry of those few reporters who had heard news of the visit.[28] Much of the substance if not the form of MacArthur's wartime approach to public affairs was subsequently maintained during the occupation of Japan.

It was not until 6 October 1945 that wartime censorship practices officially ended.[29] A run of stories issued after the end of official censorship were critical of MacArthur's headquarters handling of the press during wartime. A week later, Brigadier General Legrande Diller issued a new ruling establishing a censorship, which set up specific ceilings for the number of reporters the wire services and a mere seven major newspapers could have in the Far East zones under his headquarters' authority. The total for the entire zone, which included the Philippines, Japan, and South Korea, was 132 reporters. The three major wire services were permitted five reporters each in Japan, and newspapers like the *New York Times* were authorized to have three correspondents.[30] Because of protests from the reporters, the War Department soon suggested that MacArthur review his public relations policy. As a result, MacArthur reassigned Diller and the quota system was dropped.[31] Nonetheless, the occupation authorities retained leverage over the press. For the first several years of the occupation, the reporters in Japan depended on housing, travel, and rations dispensed by the "largesse" of MacArthur's headquarters.[32] As a result of these policies, military-media relations in the Far East were already strained well before the start of the Korean War.

Korean War

The outbreak of the Korean War and the subsequent intervention by United Nations (UN) forces led to a difficult limited war situation that came to be perceived far differently than US involvement in World War I and World War II. The challenges posed by limited war made selling the Korean War a considerable challenge for the Truman administration.[33] This was the case after several months of highly fluid fighting, when the war became stalemated in a protracted series of vicious engagements roughly along the prewar boundary between North and South Korea. General MacArthur's well-established habits of dealing with the media played an important role in determining the initial course of military-media relations in the Korean War.

On 28 June 1950, 3 days after the outbreak of the Korean War, MacArthur had decided to fly to Korea to appraise the situation on the ground. MacArthur summoned four influential press figures to his office in the Dai Ichi building. The four reporters were Russell Brines of the Associated Press (AP), Ernest Hoberecht of the United Press (UP), Howard Handleman of the International News Service, and Roy McCartney of Reuters. These men were all bureau chiefs for their respective news services in Japan. They were invited to accompany MacArthur on his visit to Korea.[34] MacArthur's party met the next day with South Korean President Syngman Rhee; Brigadier General John H. Church, head of a 15-man survey party MacArthur had dispatched to Korea on 27 June; and a number of officers from the South Korean Army and the US Military Advisory Group. After attending a briefing in a schoolhouse in Suwon, MacArthur's party traveled north to observe the front, then at the Han River south of Seoul. Along the way they observed a steady stream of South Korean soldiers retreating south. Most of the troops retained their personal weapons and carried ammunition, and most stopped to stand at attention as the three cars of MacArthur's party moved north to observe the situation. After having been on the ground for a little over 8 hours, MacArthur would return to Tokyo convinced that the defensive potential of the South Korean Army had already been exhausted; only the commitment of American ground forces could stem the North Korean advance.[35] When Brines dispatched his report to AP, he wrote:

> General MacArthur said the hundreds of South Korean soldiers seen along the road seemed to be in good shape and their morale appeared undiminished. Many saluted briskly. Most of them cheered or sang as the General's caravan passed. . . . None on a convoy of twelve trucks

> returning to the front seemed downhearted. They all cheerfully waved flags. Even most of the hundreds of tired, discouraged refugees trudging southward along the highway stopped and applauded the general.[36]

This was, of course, about as favorable a spin as one could hope to put on the rapidly deteriorating situation in Korea, where "only 24,000 ROK [Republic of Korea] troops could be located from an army that had numbered 98,000 four days earlier."[37]

On 5 July, elements of the 24th Division, designated Task Force *Smith*, became the first of the US ground forces committed to action. This piecemeal force, made up of two rifle companies and a 105-mm howitzer battery, was overrun in its first day in combat.[38] Given the chaotic conditions on the ground, it was not yet practical for any sort of formal supervision of the American reporters that were flocking to Korea. This led to a situation where reporters generally practiced self-censorship, aware that anything they got home that was perceived by MacArthur's headquarters in Tokyo as providing "aid and comfort to the enemy" was likely to lead to a loss of accreditation.[39] In any case, MacArthur, as he had in World War II, preferred to shape the news through the issue of official communiqués from his headquarters. But by the end of July, nearly 200 reporters had converged on Korea. The war they witnessed in the field was much bloodier, muddier, and frustrating than the official communiqués suggested. On 2 July, Colonel Marion Echols, a press officer in Tokyo, issued an official statement on General MacArthur's press policy. In the statement, Echols informed the press that MacArthur abhorred censorship, and no formal system as was applied in World War II would be used in Korea. Nonetheless, reporters were reminded that "inaccurate and irresponsible reporting endangers our interests and the lives of our soldiers, sailors, and airmen."[40]

To prevent the need for a formal censorship policy, reporters were told to continue to practice self-censorship. This meant they were not to report "specific units, sizes, titles, places of landings, locations, and troop movements," nor the "vilification of our armed forces personnel."[41] Echols made it clear that early reports with "emphasis to South Korea's swift collapse, had been 'far from desirable.'"[42] Ten days later, MacArthur cabled the Pentagon to complain that "voluntary press censorship has not been entirely satisfactory due to the insensate desire for sensationalism."[43] Despite his dissatisfaction, MacArthur remained committed to the principal of voluntary self-censorship for several months. He thought that such problems as there were could be rectified if the Pentagon would hold a conference with high-level press figures and better impress on them the requirements of self-censorship.[44]

In mid-July, three reporters were expelled in short order from Korea. The first was Margueritte Higgins of the *New York Herald Tribune*. Next came AP reporter Tom Lambert and UP reporter Peter Kalischer. Lambert was expelled for quoting a soldier who referred to the fighting as a "damn useless war." Kalischer was expelled because of his sensationalist stories that "made the army look bad." Higgins faced removal for the more mundane reason that she was a woman, and the US Eighth Army commander, General Walton Walker, felt that her presence was unacceptable in the crude all-male environment at the front and at Eighth Army headquarters.[45] MacArthur, facing a military-media publicity crisis, rescinded all three expulsions. In discussions with Lambert, Kalischer, and their respective bureau chiefs at the Dai Ichi on 16 July, MacArthur was informed that the wire services preferred more explicit guidance on appropriate stories from the Army, even if this meant the imposition of censorship.[46] Meanwhile in Washington, the Pentagon was being told much the same by senior representatives of several news outlets. Army Chief of Staff J. Lawton Collins and the Director of Information, Major General Floyd Parks, both thought that the Army needed to provide more specific guidance to the press than MacArthur had to date. Repeatedly in July and August, they pressed MacArthur to impose a formal censorship policy, only to be rebuffed or ignored.[47] It was only after the severe reverses suffered in December and the intervention of the Communist Chinese forces that MacArthur would accede to these requests.

Poor public relations at the Pentagon contributed to the war's first bureaucratic casualty. In August, there was even discussion of whether or not something akin to the Office of War Information should be revived. Ultimately, this project fizzled because officials in the Truman administration felt congressional and public opinion would oppose the creation of an information agency, which hitherto had only functioned during periods of total wartime mobilization. Before the outbreak of the war, Secretary of Defense Johnson had been charged with enforcing the administration's austere defense budgets. Reports of American unpreparedness in the early months of fighting, however, made it look like Johnson's cost cutting had gone too far. Johnson's own penchant for irritating fellow members of the Cabinet, his grating manner with members of key congressional committees, and several public relations missteps led President Truman to seek his resignation in early September. Truman replaced Johnson with George C. Marshall. Though some right-wing Republicans blamed Marshall for his supposed role in the "loss of China," Truman felt that Marshall was the best man to supervise the defense establishment and the defense build-up that was beginning. The appointment immediately improved relations between the State and Defense Departments and brought Marshall's considerable experience to bear in dealing with Congress and public affairs. Despite the change in personnel, the administration's handling of the war continued to face an uphill battle with congressional and public opinion.

MacArthur's decision to gamble on and the subsequent success of the landing at Inchon, of course, brought a significant shift in public perception of the war. In the intervening months, the successful UN offensive following Inchon mollified tension in military-media relations. Though some irritants between the correspondents in Korea and MacArthur's headquarters in Tokyo remained, much of the domestic news was more positive. This in turn created some concern in Washington over whether or not the public might grow too blasé, endangering the rearmament drive that was then gathering momentum.[48] But, the pendulum could and did swing both ways. Following Chinese intervention in the Korean War in late November, a much gloomier view of the war emerged that more accurately reflected the situation on the ground.[49] Though officials both in Washington and Tokyo were themselves deeply concerned, there was more concern than ever to shape public attitudes. MacArthur, recently the recipient of praise for his bold strategy, now faced the prospect of a humiliating defeat sullying his long and distinguished career. During December, MacArthur's headquarters moved closer toward a censorship policy. At the same time, a number of pronouncements by MacArthur and Major General Charles A. Willoughby, his chief of intelligence, about the massive Communist Chinese forces faced by the UN forces raised concern in Washington that the war was being portrayed as unwinnable. To assert closer control over pronouncements by "officials in the field as well as those in Washington," President Truman issued a memorandum on 5 December requiring all speeches, public statements, and press releases on foreign and military policy to be cleared by the State Department and Defense Department, respectively.[50]

In Tokyo, MacArthur's headquarters responded to reverses in the field by attempting to curtail information. On 4 December 1950, the public information officer's daily press briefings were scaled back from 60 minutes to 15 minutes. Then, on 9 December, they were temporarily suspended altogether.[51] On 18 December, Secretary Marshall and senior Pentagon officials in Washington held discussions with 11 senior press representatives who reiterated the position they had consistently supported. They were united in the view that the military was responsible for establishing security of information, and signaled their views to MacArthur in a joint communiqué. MacArthur at last agreed to institute a formal censorship program. It was announced on 22 December and went into effect 4 days later. On that date a Press Security Division would be activated within the Eighth Army, with three censors at headquarters in Taegu directed by Major Melvin B. Vorhees and two other censors working closer to the front.[52] The work of these censors would be supported by a Press Advisory Division in Tokyo under the direction of Colonel E.C. Buckhart.[53]

Though the press had long clamored for a more formal practice, this did not mean they were well disposed toward a heavy-handed censorship policy. Given the conditions in the field, this is exactly what they experienced in the weeks after the imposition of formal censorship. When General Walker was killed in a jeep accident, however, he was replaced by Lieutenant General Matthew Ridgway, who not only dramatically improved morale in Eighth Army, but also soothed the ruffled feathers of the press.[54] Ridgway retained the tough censorship code recently put into effect, but also made himself more available to reporters. On taking command of Eighth Army, Ridgway immediately set out on an inspection trip of his entire front. During this tour, he spoke openly with reporters, unafraid to speak directly about the harsh conditions of the war. His charismatic determination to improve the situation on the ground influenced reporters as well as soldiers.

Ridgway explicitly recognized the need to improve relations with the media, not just for his own sake, but to improve morale in Korea and at home. To this end, he summoned James T. Quirk to Korea. Quirk had served in North Africa and Europe as a public relations officer under Generals Omar Bradley and George Patton.[55] At the time he was recalled to Active Duty, Quirk had been working as a deputy promotion manager for the *Philadelphia Enquirer*.[56] After a marathon series of flights to Korea, which took 4 days in all, Quirk had his first meeting with Ridgway on 30 January 1950. At this meeting, Ridgway informed Quirk that he was to serve as his personal public relations advisor. After Quirk had a few days to visit the front, Ridgway sent him the following terse memo:

> MEMO FOR COL QUIRK:
>
> There are two topics I wanted to discuss with you today, and on which I would appreciate your counsel.
>
> First, how do we go about developing in the minds of our splendid men, recognition, of the almost certainty, that

> America will not in their life-time know again, the quiet, peaceful insulated comfort of other days?
>
> Second, what information can we assemble, and how and to whom transmit, so that our Government can deflate the falsely acquired military reputation of Communist China's Armies in the minds of not only of our people, but above all of the populations of Asia?
>
> Please think these over and let me have your ideas, when convenient.
>
> RIDGWAY[57]

One of the early shifts in military-media relations that Ridgway and Quirk made was to restore substantive briefings to the press. With formal censorship established, reinforced by additional public information officers dispatched by the Pentagon, the press now received much fuller briefings than MacArthur's command in Tokyo had provided in some months. Quirk, in particular, stressed to the reporters the need to understand military maneuvers in context. Hence, not every advance should be seen as a major offensive, nor should every retreat be seen as a rout. Bringing the press back into the command's confidence also allowed the reporters a clear sense of when they needed to withhold information to protect operational secrecy; and, at the same time, it allowed them to be positioned to best cover upcoming operations.[58]

Ridgway's shift from a confrontational to a cooperative stance closely paralleled the policy that Eisenhower had successfully followed with the press in Operation HUSKY during World War II. Despite Ridgway's own more positive attitude toward engagement with the press and Quirk's efforts to improve day-to-day press liaisons, there was a distinct limit to how much a local improvement in military-media relations could do within the constraints of a limited war situation. This problem was to become even more pronounced during the Vietnam War, ultimately causing what one careful observer has referred to as the "Great Divorce" between the military and the media.[59]

Vietnam

The nature of the war in Vietnam would create for the US Government and military a more difficult information environment than had been faced in the Korean War. The Korean War began with a major North Korean offensive across the 38th Parallel. Given the heightened Cold War climate and the fact that the UN sanctioned the defense of South Korea against

North Korean aggression, the Korean War, at least initially, had wider international and domestic public support. American involvement in Vietnam, certainly the method by which it was portrayed to the press, had a more ambiguous element to it.[60] South Korea's internal politics, though not without exception, were also decidedly less complicated than those of South Vietnam. When US involvement in South Vietnam was shifting from an advisory role to that of wide-ranging commitment, the South Vietnamese Government suffered two coups in close succession that cast doubts in the minds of many about the stability and ability of the South Vietnamese to prosecute the war against the Viet Cong.[61] In addition, television would bring the Vietnam War into America's living rooms in a way that had not occurred in the Korean War. Thus, from the beginning, it must be recognized that Army public affairs officers were operating in a difficult environment.

The old adage has it that generals always go about trying to refight the last war. If that were the case, the basic approach taken toward the media in Vietnam might have been rather different. As was seen in the preceding section, for the first several months of the war, General MacArthur advocated that the press follow a policy of voluntary censorship. This policy had contributed to very poor military-media relations early in the Korean War. It was only after the failures of MacArthur's policy seemed imminently apparent that a formal censorship policy was instituted in the theater. Thus, one might expect that the lesson to be learned from the Korean experience was not to dally with the imposition of censorship. However, as was the case in Korea, the US military in Vietnam repeated the precedent established by MacArthur in Korea and again rejected formal censorship in favor of a system of voluntary guidelines.[62]

However, while there was never formal censorship of journalists in the field, as had been the case in previous wars of the 20th century, there were consistent attempts at information management, which would lead to considerable ill will between the Government and the media that long-persisted after the end of the Vietnam War. In the fall of 1961, President John F. Kennedy and his senior advisors were reviewing the situation in South Vietnam in the wake of persistent reports that Ngo Dinh Diem's regime was loosing ground to the Viet Cong Communist insurgency. After a month-long review, which included a careful assessment of the situation in Vietnam by General Maxwell Taylor, National Security Council (NSC) staff member Walt Rostow, and others, President Kennedy approved National Security Action Memorandum (NSAM) 111 on 22 November 1961.[63] This authorized the US Government to undertake a number of measures to strengthen its joint effort with the Government of Vietnam (GVN) to arrest the deterioration of the situation. This would include the augmentation of the Military Assistance Advisory Group (MAAG) with increased airlift for GVN forces. These additional airlift units were to be flown by US personnel. Though the NSAM did not specify the number of troops, there was a clear trend toward a considerable expansion in the US advisory presence in Vietnam. In return for increased US assistance, Ngo Dinh Diem's regime was to undertake a number of reforms to improve the GVN's capability for prosecuting the war.[64] President Kennedy, however, desired that this expansion in US effort not lead to a perception that the United States was going to replace or supplant the Army of the Republic of Vietnam (ARVN), which was to retain primary responsibility for combating the insurgency.

Around the same time as President Kennedy approved NSAM 111, US officials in South Vietnam were instructed to refrain from providing the press with any information regarding military or political activities; instead, inquiries were to be directed to South Vietnamese officials. As US assistance to the GVN increased in the following year, Commander in Chief Pacific (CINCPAC) Admiral Harry Felt issued orders that no reporters were to accompany US helicopters transporting ARVN forces and their American advisors on combat missions.[65] To clarify the US position, a joint State–Defense–United States Information Agency (USIA) message was sent to the US Embassy in Saigon on 21 February 1962. While ostensibly aimed at increasing the ability of US officials to deal with the press, "the cable prompted the U.S. mission in Saigon to persist in the practice of excessive classification to a degree that denied newsmen access to whole segments of the war."[66]

During the next 2 years, while officially American involvement was strictly advisory, there was a concerted effort among the upper-echelons of US officials in Vietnam to portray South Vietnamese efforts in as positive a light as possible. Criticism of South Vietnamese shortcomings, from the perspective of the Kennedy administration, would only serve to undermine public support in the United States for continued aid to South Vietnam. Further, it would complicate relations between Washington and Saigon, and serve ready-made propaganda to the Communists world to use against Diem's regime. However, the reporters in the field, who saw the war at a more localized level, were increasingly skeptical of what they saw as discord between the war on the ground and its official portrayal.[67] When the Diem regime moved to expel several reporters, including *New York Times* correspondent Homer Bigart, *Newsweek* correspondent Francois

Sully, and NBC News correspondent James Robinson, who had written stories or made comments unflattering of the regime's handling of the war, it exacerbated relations between the press corps and the South Vietnamese Government.[68]

When elements of the ARVN's 7th Division badly mismanaged an operation against the 514th Viet Cong Battalion in early January 1963 near the village of Ap Bac in Dinh Tuong province, it provided a focal point for the Western press to vent frustration with the Diem regime. A number of American advisors were killed in the action, and five US helicopters were lost through a combination of mechanical failures and enemy fire. Though no reporters had been present at the first day of the battle, they were soon able to piece together the situation through interviews with frustrated US Army advisors on the scene who had seen the superior forces of the 7th Division allow the smaller 514th Battalion slip out of what should have been a decisive defeat. Other than the disgruntled US advisors, the US mission officially tried to avoid any critical commentary on the action at Ap Bac. Nonetheless, the story quickly became a focal point in the United States regarding the South Vietnamese Government's handling of the war in general.[69] According to William Hammond, "Ap Bac and the controversy surrounding it marked a divide in the history of U.S. relations with the news media in South Vietnam" because, "after it, correspondents became convinced that they were being lied to and withdrew, embittered, into their own community."[70] But the nadir was yet to come.

During the summer of 1963, confidence in Diem's leadership plummeted further when a number of Buddhist monks committed suicide by self-immolating themselves in public as a means of protest. By year's end, sufficient unrest in South Vietnam led to Diem's overthrow by a military coup, tacitly approved by the US Government. Though the removal of the unpopular Diem would temporarily leaven criticism of the war effort, the instability of the military leaders who followed in Diem's wake hardly aided the attempts of the US Government to portray progress in the stabilization of South Vietnam.

In the spring of 1964, the US Government moved to impose closer information coordination on its efforts in South Vietnam. In the Pentagon, Colonel Rodger Bankson, an experienced public affairs officer who had worked as a censor during the Korean War, was appointed to lead the newly created Southeast Asia Division within the Office of the Assistant Secretary of Defense for Public Affairs. In South Vietnam, an Office of Information within the Military Assistance Command, Vietnam (MACV) was established. President Lyndon B. Johnson, acting on a recommendation from US Information Agency director Carl Rowan, appointed Barry Zorthian to serve as the chief public affairs officer.[71] Zorthian was a retired Marine who had served in World War II, worked as a journalist, and served with Voice of America. Zorthian originally was recruited to serve in Vietnam as the public relations officer for Voice of America. His writ would soon be considerably expanded, however. At a high-level conference in Honolulu in early June, there was widespread agreement of the need to appoint a single communications "czar" for the US military and civilian missions in South Vietnam.[72]

In the summer of 1964, the US mission in Vietnam underwent a number of changes in key personnel. General Maxwell Taylor, Chairman of the Joint Chiefs of Staff, would replace Henry Cabot Lodge as US Ambassador in Vietnam; General William Westmoreland replaced General Paul Harkins as Commander, MACV; and Barry Zorthian assumed his duties as the chief public affairs officer. Zorthian and Westmoreland did everything possible to make the MACV Office of Information the central source for information on military operations.[73] The State Department issued instructions that the public information program was to stress "maximum candor."

The program of "maximum candor," however, was undermined little more than a year after it was put in place. In April 1965, President Johnson approved a range of US actions in Vietnam that included increased US military operations. To make it appear as though there had not been a change in policy that deliberately escalated the war, NSAM 328 instructed Government officials:

> The actions themselves should be taken as rapidly as practicable, but in ways that should minimize any appearance of sudden changes in policy, and official statements on these troop movements will be made only with the direct approval of the Secretary of Defense, in consultation with the Secretary of State. The President's desire is that these movements and changes should be understood as being gradual and wholly consistent with existing policy.[74]

American public support for the war had risen considerably in response to the Gulf of Tonkin incident. On 13 February 1965, President Johnson authorized Operation ROLLING THUNDER, a sustained bombing campaign against North Vietnam that would continue, with intermittent bombing "pauses" until near the end of Johnson's presidency. Initially, the Gallup Poll reported a 67 percent public approval rating of the campaign, though telegrams addressed to the White House ran 14 to 1 against the new policy.[75] Press opinion in Vietnam, which had been deteriorating steadily since late 1963, had been very negative about the situation in country prior to this most recent escalation of the war. Editorial opinion in the United States, conversely, was still largely supportive of the administration's policies. In mid-May, *Time* magazine ran a story on the war titled, "Viet Nam: The Right War at the Right Time."[76] Nonetheless, dissent was spreading beyond just members of the press corps in South Vietnam. Around the same time that *Time* carried its upbeat article, John Mecklin published a critical book on the US effort in Vietnam.[77] Although there were others, Mecklin's critique was one of the first of the dissents to emerge from the "inside."[78]

Despite increasing skepticism, the overall American public remained supportive of the war effort from 1965 to 1967. But as the number of troops deployed and the number of US casualties continued to mount, so did criticism of the war. In November 1967, Westmoreland was recalled to the United States as part of a concerted effort by President Johnson to shore up eroding public and Congressional support for the war. During Westmoreland's visit, he testified before Congress, spoke on *Meet the Press*, and addressed the National Press Club. In his widely reported remarks, Westmoreland argued that the war had reached a turning point, and that he foresaw the possibility of winding down US troop commitments within 2 years. He also stated his belief that the Viet Cong were no longer capable of mounting large-unit actions in South Vietnam. Then, 2 months later, the Viet Cong and infiltrated units of the North Korean Army launched the Tet Offensive.[79] The Tet Offensive marked a fundamental break in the *perception* of the war. Though the Tet Offensive broke the back of the Viet Cong and was ultimately a military victory for the US and ARVN forces, the scale of the assault and the operational surprise achieved belied the positive assessments that Westmoreland and others had cultivated. In many ways, the political consequences the Tet Offensive produced did ultimately undermine the prospect of a long-term US commitment to the defense of South Vietnam. From the perspective of military-media relations, Tet cast a long shadow on an entire generation of officers. The lieutenants, captains, and majors who served in the Army during the Tet Offensive came to believe that it was the media's portrayal of Tet that turned a military victory into defeat.[80] Many of the problems that continued to plague military-media relations through the first Gulf War were a result of this perception.

Grenada

In the aftermath of the American involvement in Vietnam, military-media relations reached a point from which the relationship was long in recovering. Not only had the military and the media come to distrust one another, but American society as a whole suffered through a long period of distrust of public institutions. Vietnam, racial and social discord, the Watergate scandal, and the hearings conducted by Senator Frank Church (Democrat, Idaho) on covert operations all contributed to the erosion of trust between the American people and public institutions. The long hostage crisis in Iran at the end of the Carter administration and the aborted rescue attempt, of course, did little to re-instill the confidence of the American populace or the press in the effectiveness of the guardians of national security. According to Vice Admiral Joseph Metcalf III, who served as commander of the 1983 intervention in Grenada (Operation URGENT FURY): "At the time of the Grenada intervention, relations between the press and the US military had been eroded to an appalling state. . . . The military brooded over the loss in Vietnam, and many blamed the press. At the same time, the media was deeply suspicious of those in authority within the military and its surrogate, the Pentagon."[81] As a consequence of these deeply held suspicions, Metcalf described senior military and civilian officers at the Pentagon in the late 1970s and 1980s as being more pre-occupied to reacting to the Early Bird (an in-house Department of Defense synopsis of leading news articles pertaining to military matters) than "attending to the business of the Cold War."[82] When the Reagan administration chose to intervene on the island of Grenada in October 1983, the poor relations between the military and the media were reflected in the decision not to notify any members of the media before the invasion took place and to exclude them from landing on the island for 2 days after the commencement of operations.[83] According to Metcalf, who had only 39 hours notification that he was to lead the operation prior to the scheduled first landing of troops, when he was informed by a Public Affairs Officer sent by the Commander in Chief Atlantic to brief him on press policies, he was told that there would be no press involvement in the operation. For his part, Metcalf was unconcerned with this decision.[84]

Though Operation URGENT FURY was a relatively minor military operation in the broad sweep of the 20th century, the media's exclusion generated considerable reaction. Chairman of the Joint Chiefs General John H. Vessey Jr. was sufficiently concerned with the volume of media agitation over being excluded from the early operation that he asked Winant Sidle, a retired brigadier general who had served as the MACV Chief of Information in 1967–68, to chair a commission on military-media relations. Skirting the issue of censorship, the Sidle Commission recommended the creation of a National Press Pool. The National Press Pool would have reporters available for deployment with military units on short-notice to cover operations like URGENT FURY.[85] The pool was approved by the Joint Chiefs and the Secretary of Defense in April 1985. On 10 occasions over the next 4 years, members of the pool deployed with military forces on exercises. When they deployed to Panama during Operation JUST CAUSE in 1989, however, they found themselves escorted to a room where they watched CNN coverage of the war and then were treated to a State Department lecture on Panamanian history. As had been the case in Grenada, it was several days before journalists were allowed into the field, by which time most of the fighting was already over.[86]

Although members of the National Press Pool again deployed with American forces to Saudi Arabia during Operation DESERT SHIELD, as it turned out there was a long period after the initial deployments before much of note happened in the field. As the mission shifted from the defense of Saudi Arabia toward a potential operation to eject Iraqi forces from Kuwait, the initial media pool swelled dramatically in size. At the same time, US Central Command (CENTCOM) had plenty of time to develop guidelines for the numerous members of the press who clamored to cover the confrontation between the United States-forged coalition and Iraq.[87] From the US military's perspective, the first Gulf War was a rather successful example of military-media engagement.[88] Indeed, DESERT STORM proved to be a big story, but one in which the media perceived itself as having been carefully managed by the Pentagon, seemingly still smarting from its own perceptions of the media's impact on the Vietnam War. After Operation DESERT STORM, there would be another bout of analysis about how to improve military-media relations that would produce reams of congressional testimony and numerous books.[89] By the second Gulf War, the US military had adopted a policy that would prove more conducive to the media's desire to be close to the action by resurrecting the practice of imbedding reporters with combat units.

From the First Gulf War to the Global War on Terrorism

After the first Gulf War, Alan D. Campen, a retired Air Force Colonel who had served as Director, Command and Control Policy, Office of the Under Secretary of Defense for Policy from 1982 to 1985, edited a volume on the war titled, *The First Information War: The Story of Communications, Computers, and Intelligence Systems in the Persian Gulf War* (1992). The book opens by stating, "The United States unveiled a radically new form of warfare in the Persian Gulf in 1991." This "radically new form of warfare" was brought about by leveraging information.[90] The interest in leveraging information grew increasingly important in the 1990s, when the real-time coverage of overseas conflicts, dubbed the "CNN effect," became commonplace. A study group at the Center for Strategic and International Studies at the National War College produced a report in March 1993 that looked at the threat environment of the 1990s and attempted to delineate the criteria for US military forces in this period. Among a number of other points, it recognized the need to "fight a CNN war" in which the US armed forces:

> Must be capable of responding to media demands for instantaneous information, and of using the rapid transmission of data to its advantage. This magnifies the importance of tending to image considerations, the first criterion, especially in terms of the friendly fire problem. But it also suggests the need for greater information dominance and for some thought about how modern, real-time news reporting can be used to U.S. advantage in future military operations.[91]

The impact of the footage of a US Army helicopter pilot being dragged through the streets of Mogadishu in October 1993 during the UN peacekeeping mission there dramatically illustrated the influence of the "CNN effect."[92] By mid-decade, however, some observers suggested a coming backlash against the "CNN effect." Marine Commandant General Charles Krulak argued in the fall of 1996 that there was a "growing reluctance to fight for CNN—the concern that we are being manipulated into conflict by media interests rather than legitimate interests."[93] This idea that the global media was determining rather than merely reporting the headlines, in a manner not unlike the jingoistic press in America before the Spanish-American War, had enough public currency by the mid-1990s to serve as the plot of the James Bond film *Tomorrow Never Dies* (1997).

Ironically, the proliferation of the very technology that made the "CNN effect" possible in the 1990s has undermined the ability of any particular media outlet to retain the ability to dominate the contemporary news agenda. In the 21st century, it seems unlikely that either governments or specific media outlets will be able to dictate the flow and dissemination of information. The proliferation of satellite telecommunications, the internet, and cheap digital recording and communication devices have all undermined the ability of any government or media outlet to manage information. At the same time, the US Government and military have continued to pursue improved modes of information management. In August 1998, President William J. Clinton authorized Tomahawk cruise missile strikes on targets in Sudan and Afghanistan linked to Osama bin Laden's terrorist network in reprisal for the attacks on two US Embassies in East Africa. The rapidity with which the Sudanese Government launched its own information campaign to discredit the US cruise missile strike surprised the Clinton administration.[94] As a result, President Clinton approved Presidential Decision Directive 68 in April 1999 to encourage a better governmental approach to information management. Though this directive remains classified, among other things it authorized the establishment of an International Public Information System and an International Public Information Core Group chaired by the Under Secretary of State for Public Diplomacy and Public Affairs, whose charter was reminiscent of the Truman administration's Psychological Strategy Board.[95]

After the terrorist attacks of 11 September 2001, some within the Bush administration argued that the US Government's capabilities for information management were insufficient for the ideological dimension of the conflict with al-Qaeda. This seems to have been the case within Secretary of Defense Donald Rumsfeld's department. In October 2001, Under Secretary of Defense for Policy Douglas Feith set up an Office of Strategic Influence (OSI) in the Pentagon.[96] The OSI proved to be rather short-lived, however, when it came under intense public scrutiny for purportedly making plans to disseminate false news stories to foreign media in early 2002. Feith dissolved the OSI at the end of February 2002.[97] Slightly less than a year later, President George W. Bush created an Office of Global Communications in the White House, headed by the Deputy Assistant to the President for Global Communications.[98] During the remainder of the Bush presidency, the importance of strategic communications to the Global War On Terrorism (subsequently the Long War) was frequently reiterated.[99] The 2006 Quadrennial Defense Review, for instance, stated:

> Victory in the long war ultimately depends on strategic communication by the United States and its international partners. Effective communication must build and maintain credibility and trust with friends and foes alike, through an emphasis on consistency, veracity and transparency both in words and deeds. Such credibility is essential to building trusted networks that counter ideological support for terrorism.[100]

Yet, for all the interest in strategic communications during the Bush administration, it may well be that the US Government's approach to information management in the early years of the 21st century largely mirrored the patterns of information management of the preceding century.

Notes

1. Dwight David Eisenhower, *At Ease: Stories I Tell to Friends* (Garden City, NY: Doubleday & Company, 1967), 320–321.

2. Kenneth Osgood, *Total Cold War: Eisenhower's Secret Propaganda Battle at Home and Abroad* (Lawrence, KS: University Press of Kansas, 2006), 49. For the best account of the US Army's approach to psychological warfare, see Alfred H. Paddock Jr., *U.S. Army Special Warfare: Its Origins*, rev. ed. (Lawrence, KS: University Press of Kansas, 2002).

3. J. Lawton Collins, *Lightning Joe: An Autobiography* (Baton Rouge, LA: Louisiana State University Press, 1979), 339–341.

4. J. Lawton Collins, "An Information Policy for the New Army," *Army Information Digest* 1, No. 1 (May 1946): 3–5.

5. Collins, *Lightning Joe*, 343.

6. Ibid., 344.

7. Survey of the Operations of the Bureau of Public Relations of the War Department, transmitted to Major General Floyd Park by Jack H. Lockhart, 3 June 1946. Papers of Floyd H. Parks, Box 10, Dwight D. Eisenhower Library.

8. Ibid., part I, 2.

9. Ibid., part I, 3.

10. Ibid.

11. Ibid, part I, 6.

12. Ibid, part II, 2–3.

13. Ibid, part II, 6.

14. Ibid., part II, 6–7.

15. Ibid., part I, 4.

16. In 1951, the Army Information School became the Armed Forces Information School, though it generally was still referred to as the Army Information School. In 1964, it became the Defense Information School. The school is currently located at Fort Meade, Georgia. "Defense Information School Training." http://www.usmilitary.com/2163/defense-information-school-training/ (accessed 14 May 2009).

17. "The Army Information School Makes a Bow," *Army Information Digest* 1, No.1 (May 1946): 14–19.

18. Doris Condit, *History of the Office of the Secretary of Defense [HOSD], Vol. II: The Test of War 1950–1953* (Washington, DC: Historical Office, Office of the Secretary of Defense, 1988), 26.

19. Condit, *HOSD*, *Vol. II,* 26–27.

20. Prior to the Korean War, there was a long discussion within the US Government of the appropriate roll and scope of public information campaigns, both of an overt and covert nature. This subject is beyond the scope of this study; however, a fairly extensive literature on this subject has developed. See Nancy E. Bernhard, *U.S. Television News and Cold War Propaganda, 1947–1960* (New York, NY: Cambridge University Press, 1999); Sarah-Jane Corke, *US Covert Operations and Cold War Strategy: Truman, Secret Warfare, and the CIA, 1945–53*

(London: Routledge, 2008); Robert T. Davis II, "Cold War Interagency Relations and the Struggle to Coordinate Psychological Strategy," in Kendall D. Gott and Michael G. Brooks, eds., *The US Army and the Interagency Process: Historical Perspectives* (Fort Leavenworth, KS: Combat Studies Institute Press, 2009), 301–319; Peter Grose, *Operation Rollback: America's Secret War behind the Iron Curtain* (Boston, MA: Houghton Mifflin, 2000); Walter Hixson, *Parting the Curtain: Propaganda, Culture and the Cold War, 1945–1961* (New York, NY: St. Martin's Press, 1997); David F. Krugler, *The Voice of America and the Domestic Propaganda Battles, 1945–1953* (Columbia, MO: University of Missouri Press, 2000); Scott Lucas, *Freedom's War: The American Crusade against the Soviet Union* (New York, NY: New York University Press, 1999); Gregory Mitrovich, *Undermining the Kremlin: America's Strategy to Subvert the Soviet Bloc, 1947–1956* (Ithaca, NY: Cornell University Press, 2000); Michael Nelson, *War of the Black Heavens: The Battles of Western Broadcasting in the Cold War* (Syracuse, NY: Syracuse University Press, 1997); Frank Ninkovich, *U.S. Information Policy and Cultural Diplomacy* (Washington, DC: Foreign Policy Association, 1996); and Thomas C. Sorensen, *The Word War: The Story of American Propaganda* (New York, NY: Harper & Row, 1968).

21. Geoffrey Perret, *Old Soldiers Never Die: The Life of Douglas MacArthur* (New York, NY: Random House, 1996), 74.

22. D. Clayton James, *The Years of MacArthur*, *Vol. II: 1941–1945* (Boston, MA: Houghton Mifflin, 1975), 90.

23. Ibid., *Vol. II,* 164–165.

24. Ibid., *Vol. II*, 280–281.

25. Ibid., *Vol. III*, 11; and Perret, *Old Soldiers Never Die*, 473.

26. Luther J. Reed, "Tokyo Reporting," *Army Information Digest* 1, No. 4 (August 1946): 17–19.

27. James, *The Years of MacArthur*, *Vol. III;* and Perret, *Old Soldiers Never Die*, 17–22, 497–498.

28. William J. Coughlin, *Conquered Press: The MacArthur Era in Japanese Journalism* (Palo Alto, CA: Pacific Books), 116.

29. Ibid., 117.

30. Ibid., 118–119.

31. Ibid., 119.

32. Perret, *Old Soldiers Never Die*, 528–529.

33. For the best account of this challenge, see Steven Casey, *Selling the Korean War: Propaganda, Politics, and Public Opinion in the United States, 1950–1953* (Oxford, NY: Oxford University Press, 2008).

34. Stanley Weintraub, *MacArthur's War: Korea and the Undoing of an American Hero* (New York, NY: The Free Press, 2000), 48.

35. James, *The Years of MacArthur*, *Vol. III*, 426–427. MacArthur's report on the trip to the Joint Chiefs of Staff and State Department is reproduced in United States Department of State, *Foreign Relations of the United States [FRUS] 1950, Vol. VII* (Washington, DC: US Government Printing Office, 1950), 248–250.

36. Quoted in Weintraub, *MacArthur's War*, 54.

37. James, *The Years of MacArthur*, *Vol. III,* 425.

38. Roy K. Flint, "Task Force Smith and the 24th Division: Delay and Withdrawal, 5–19 July 1950," in *America's First Battles*, ed. Charles E. Heller and William A. Stofft (Lawrence, KS: University Press of Kansas, 1986), 266–299.

39. Weintraub, *MacArthur's War*, 67.

40. Casey, *Selling the Korean War*, 45.

41. Ibid.

42. Ibid., 49.

43. Quoted in Casey, *Selling the Korean War*, 53.

44. Ibid.

45. Ibid., 54–55.

46. Ibid., 55.

47. Ibid., 59–60.

48. On 19 July, President Truman had requested an additional $10 billion defense appropriation. This, combined with an earlier supplemental request, had doubled the defense budget over the level that the administration had previously established as its peacetime budgetary ceiling.

49. D. Clayton James takes a view rather more sympathetic to MacArthur's perspective, stating, "Much of the American and West European press reacted with sensational, alarmist reports on the aborted Home by Christmas Drive, portraying MacArthur as blundering and warmongering and his forces as fleeing, panic-stricken, in the face of the Chinese attackers." James, *The Years of MacArthur*, *Vol. III*, 539.

50. The directive is reproduced in Declassified Documents Reference System [DDRS] (Woodbridge, CT: Research Publications International, 1992), Appendix I, 1992, F127, 1873.

51. Casey, *Selling the Korean War*, 147.

52. Ibid., 158.

53. Ibid., 155.

54. Roy E. Appleman wrote "[Ridgway] alone made the difference in keeping the American Eighth Army in Korea, gradually turning it around to face north once again and to emerge as a strong, motivated fighting force in the Chinese 4th and 5th Phase offensives in the winter and spring of 1951." Roy E. Appleman, *Ridgway Duels for Korea* (College Station, TX: Texas A&M University Press, 1990), xiv.

55. Rory Quirk, *Wars and Peace: The Memoirs of an American Family* (Novato, CA: Presidio Press, 1999), 19, 37–38.

56. Ibid., 154.

57. Ibid., 167–168.

58. Casey, *Selling the Korean War*, 167–168.

59. This is the title of chapter 5, covering Korea and Vietnam, in Michael S. Sweeney, *The Military and the Press: An Uneasy Truce* (Evanston, IL: Northwestern University Press, 2006).

60. Syngman Rhee would serve continuously as president of South Korea from 1948 until 1960. From 1919 to 1939, he had served as president of the Korean Provisional Government-in-Exile. There was no such comparable figure in the history of South Vietnam.

61. Ngo Dinh Diem was overthrown on 1 November 1963. He and his brother, who had headed the secret police, were killed the following day.

62. Daniel C. Hallin, *The "Uncensored War": The Media and Vietnam* (New York, NY: Oxford University Press, 1986), 6; and William M. Hammond, *Reporting Vietnam: Media and Military at War* (Lawrence, KS: University Press of Kansas, 1998), 291.

63. Robert D. Schulzinger, *A Time for War: The United States and Vietnam, 1941–1975* (New York, NY: Oxford University Press, 1997), 107–113.

64. John F. Kennedy Presidential Library & Museum, National Security Act Memorandum (NSAM) 111: First Phase of Viet-NAM Program, 22 November 1961. http://www.jfklibrary.org/Historical+Resources/Archives/Reference+Desk/NSAMs.htm (accessed 14 May 2009).

65. Clarence R. Wyatt, "The Media and the Vietnam War," in *The War That Never Ends: New Perspectives on the Vietnam War*, eds. David L. Anderson and John Ernst (Lexington, KY: University Press of Kentucky, 2007), 272–273.

66. This telegram is generally referred to as State Department Cable 1006. It is reproduced as "Tel. from Dep. of State to Emb. Vietnam, 21 February 1962," *FRUS 1961–1963, Vol. II*, Doc.75. See also William M. Hammond, *Public Affairs: The Military and the Media, 1962–1968* (Washington, DC: Center of Military History, 1988), 15.

67. Hammond, *Public Affairs: The Military and the Media, 1962–1968*, 17–24.

68. Ibid., 24–29.

69. Ibid., 30–35.

70. Ibid., 37.

71. Hammond, *Reporting Vietnam*, 22–23.

72. *FRUS 1964–1968, Vol. I*, Doc. 189: Summary Record of Meetings, Honolulu, 2 June 1964.

73. Hammond, *Reporting Vietnam*, 23.

74. NSAM 328, "Presidential Decisions Regarding U.S. Policy in Vietnam, 6 April 1965." http://www.fas.org/irp/offdocs/nsam-lbj/index.html and http://www.lbjlib.utexas.edu/johnson/archives.hom/nsams/nsamhom.asp (accessed 14 May 2009).

75. Schulzinger, *A Time for War*, 172.

76. *Time*, 14 May 1965.

77. John Mecklin, *Mission in Torment: An Intimate Account of the U.S. Role in Vietnam* (Garden City, NY: Doubleday & Company, 1965).

78. Two reporters, both of whom received Pulitzer Prizes in 1964 for their coverage of Vietnam, published accounts of the Vietnam War in early 1965. See David Halberstam, *The Making of a Quagmire* (New York, NY: Random House,

1965), and Malcolm Browne, *The New Face of War* (Indianapolis, IN: Bobbs-Merrill, 1965).

79. Two good recent assessments are David F. Schmitz, *The Tet Offensive: Politics, War, and Public Opinion* (Lanham, MD: Rowman & Littlefield, 2005); and James H. Wilbanks, *The Tet Offensive: A Concise History* (New York, NY: Columbia University Press, 2007). Peter Baestrup's careful assessment of how Tet was covered by the media is also of note. Peter Braestrup, *Big Story: How the American Press and Television Reported and Interpreted the Crisis of Tet 1968 in Vietnam and Washington*, 2 vols. (Boulder, CO: Westview Press, 1977).

80. Douglas Kinnard, *The War Managers* (Hanover, NH: University of New England Press, 1977), 124–135; Sweeney, *The Military and the Press*, 151–152.

81. J. Metcalf III, "The Press and Grenada, 1983," in *Defence and the Media in Time of Limited War*, ed. Peter R. Young (London: Frank Cass, 1992), 168.

82. Ibid., 169.

83. In 1979, Maurice Bishop, leader of a Marxist-leaning party, had seized power in a bloodless coup. During the period from 1979 to 1983, Grenada had established close ties with Fidel Castro. Bishop was himself overthrown by a left-wing faction of his own government on 12 October 1983 and executed a week later. The coup leaders dissolved the civilian government, established a so-called Revolutionary Military Council, closed the airport, and imposed a curfew under which violators were subject to be shot on sight. With approximately 1,000 US citizens on the island, including around 600 medical students, the Reagan administration promptly initiated contingency planning for their evacuation. On 25 October 1983, a brigade of the 82d Airborne and US Marines landed on the Caribbean island of Grenada to prevent the seizure of the medical students as hostages and provide for the establishment of a government more amenable to American national security interests. Ronald H. Cole, *Operation Urgent Fury: The Planning and Execution of Joint Operations in Grenada 12 October–2 November 1983* (Washington, DC: Joint History Office, Office of the Chairman of the Joint Chiefs of Staff, 1997).

84. J. Metcalf III, "The Press and Grenada, 1983," 169.

85. Sweeney, *The Military and the Press*, 155–156.

86. Ibid., 157–158.

87. See the Guidelines for the Media, 14 January 1991, reproduced in appendix A.

88. Colin Powell with Joseph E. Persico, *My American Journey* (New York, NY: Random House, 1995), 528–530.

89. Senate Committee on Governmental Affairs, "Pentagon Rules on Media Access to the Persian Gulf War," *Hearing before the Committee on Governmental Affairs United States Senate*, 102d Cong., 1st sess., 20 February 1991 (Washington, DC: US Government Printing Office, 1991). See also Robert E. Denton Jr., *The Media and the Persian Gulf War* (Wesport, CT: Praeger, 1993); John J. Fialka, *Hotel Warriors: Covering the Gulf War* (Washington, DC: Woodrow Wilson Center Press, 1991); John R. MacArthur, *Second Front: Censorship and Propaganda in*

the Gulf War (New York, NY: Hill and Wang, 1992); Twentieth Century Fund, *Battle Lines: Report of the Twentieth Century Fund on the Military and the Media* (New York, NY: Priority Press, 1985).

90. Alan D. Campen, ed., *The First Information War: The Story of Communications, Computers, and Intelligence Systems in the Persian Gulf War* (Fairfax, VA: AFCEA International Press, 1992), ix.

91. Michael J. Mazaar, Jeffrey Shaffer, and Benjamin Ederington, *Military Technical Revolution: A Structural Framework* (Washington, DC: Center for Strategic and International Studies, 1993), 11–12.

92. President Clinton wrote in his memoirs that the Battle of Mogadishu was "one of the darkest days of my presidency," and that the event "haunted me." Subsequent to the ghastly images of an Army helicopter pilot being drug through the streets, the Clinton administration and Congress adopted a 6-month timeline for the withdrawal of US forces from Somalia. William J. Clinton, *My Life* (New York, NY: Alfred A. Knopf, 2004), 576. See also Frank J. Stech, "Winning CNN Wars," *Parameters* (Autumn 1994): 37–38.

93. Charles C. Krulak, "The United States Marine Corps in the 21st Century," *RUSI Journal* (August 1996): 24.

94. Jamie Frederic Metzl, "Popular Diplomacy," *Daedalus* 128, No. 2 (Spring 1999): 177–178.

95. International Public Information (IPI), Presidential Decision Directive 68, 30 April 1999. http://www.fas.org/irp/offdocs/pdd/pdd-68.htm (accessed 14 May 2009).

96. Douglas J. Feith, *War and Decision: Inside the Pentagon at the Dawn of the War on Terrorism* (New York, NY: Harper, 2008), 171.

97. Ibid., 173–176.

98. Executive Order 13282, "Establishing the Office of Global Communications, 21 January 2003," John T. Woolley and Gerhard Peters, *The American Presidency Project* [online] (Santa Barbara, CA: University of California (hosted), Gerhard Peters (database)). http://www.presidency.ucsb.edu/ws/?pid=61379 (accessed 14 May 2009).

99. There is substantial literature on strategic communications. See, for example, Jeffrey B. Jones, "Strategic Communication: A Mandate for the United States," *Joint Force Quarterly* 39 (Fourth Quarter, 2005), 108–114; Pamela Keeton and Mark McCann, "Information Operations, STRATCOM, and Public Affairs," *Military Review* 85, No. 6 (November–December 2005), 83–86; Huba Wass de Czege, "On Winning Hearts and Minds," *Army* (August 2006), 8–18; Carl D. Gunrow, "Winning the Information War," *Army* (April 2007), 12–15; *US National Strategy for Public Diplomacy and Strategic Communications* (Washington, DC: Strategic Communication and Public Diplomacy Coordinating Committee, 2007); Dennis M. Murphy and James F. White, "Propaganda: Can a Word Decide a War?" *Parameters* 37 (Autumn, 2007), 15–27; Richard Halloran, "Strategic Communication," *Parameters* 37 (Autumn, 2007), 4–14; Carnes Lord, "On the Nature of Strategic Communications," *Joint Force Quarterly* 46 (Third Quarter,

2007), 87–89; William M. Darley, "The Missing Component of U.S. Strategic Communications," *Joint Force Quarterly* 47 (Fourth Quarter, 2007), 109–113.

100. *Quadrennial Defense Review Report* (Washington, DC: Department of Defense, 6 February 2006), 91–92.

Chapter 5
Conclusions

> *For the journalist, war sells the news. But news gathered from battlefields and wartime governments also has the higher goal of fulfilling the press's modern role as the Fourth Estate. News provides American citizens with information to help them make informed decisions about their leaders. For the soldier, however, news is primarily a tool or a weapon. Information shared through the mass media can bolster military and civilian morale, raise enlistments, boost the armed forces' budget, undermine the enemy's confidence, and hasten the end of conflict. Or it can compromise battlefield security, wreck the civilian base of support, and topple the government leadership directing the war effort. From the military point of view, a reckless press can turn victory into defeat.*
>
> Michael S. Sweeney[1]

Surveying the history of military-media relations during conflicts over the last 150 years, one recognizes that there has been a rich variety of experiences. There has always been and always will be tension between military authorities and the press, which are more pronounced when the military situation is most difficult. This creates a particular problem when operations are difficult because of the military's increased desire to tighten control over information, both to prevent exploitation by the enemy and to protect the morale of its troops. Members of the media may see these same circumstances as a time when an open inquiry into the causes of such reverses should be investigated and exposed to the American public. Yet, in many instances, the military and the media have worked out reasonable accommodations that have permitted extensive news coverage of wars without creating debilitating friction between the corporate goals of either group. Though reporters have chafed at seemingly arbitrary practices and military officers have despaired at achieving "balanced" reporting, there has been no instance since General Sherman's infamous confrontation with Thomas Knox that a military commander has seen a need to court-martial a reporter. By the second Gulf War, even the considerable rift in military-media relations during the Vietnam War seems to have been bridged with the return to the time-honored practice of allowing reporters to embed themselves with combat units.

A number of factors affect military-media relations. In many cases, the activities of Army public affairs officers are only ancillary to the relationship. That is to say, military-media relations are seldom merely a matter of providing or not providing food, shelter, transportation, and communications facilities to the press. The more important determinants of the relationship are sometimes factors beyond the control of the military. When US military forces operate with allies, particularly in an allied-host country, thorny question can arise. It is not hard to imagine situations in which the performance of an allied force or forces is cause for great frustration. Nonetheless, the continued participation of that ally may have considerable political importance beyond the purview of a military officer's bailiwick. When reporters ask officers what they think of their allies, what is more important—honesty or preserving the relationship with the ally? Is the reporter's pursuit of the "truth" more important than the preservation of a military coalition? Another difficult problem for the military arises when presidential policies have deliberately precluded "maximum candor" with the press. Again, the military officer would find himself in a dilemma of following orders, which precludes full cooperation with the media. In some cases, individual Army commanders have also pursued policies that have done a good deal to undermine the Army's pursuit of positive public affairs. Notable examples include Commanding General Nelson Miles during the Philippine-American War and General Douglas MacArthur in Korea.

As for the future, changing modes in the nature of global communications may fundamentally alter the very nature of military-media relations.[2] In past situations where military control over the modes of communication between a theater and the United States was well developed, there was generally a tendency toward censorship in theater. Even when there was no formal censorship, practices aimed at controlling the information available to the press were often followed. However, the proliferation of communications methods in recent years has made theater censorship and information control increasingly less relevant. Stripped of the ability to control the dissemination of material, the US Government and the military have had to place greater stress on the shaping of the message. The renewed interest in public diplomacy and strategic communications since the events of 9/11 is symptomatic of this trend. The mission statement of the Office of Global Communications, established by President George W. Bush on 21 January 2003, clearly enunciated this interest. Executive Order 13283, which established the Office of Global Communications, states that its purpose was to advise the President and the heads of executive departments and agencies on how best to "ensure consistency in messages that will promote the interests of the United States abroad, prevent misunderstanding, build support for and among coalition partners of the United States, and inform international audiences."[3] However, when such wide-ranging information coordination organizations were created in the past, it often made it difficult to draw a clear line between public relations and propaganda and/or psychological warfare. As General William Tecumseh Sherman put it so long ago, "We do not want the truth about things; that is what we don't want . . . we do not want the enemy any better informed about what is going on here than he is."[4] This contrasts with Jack Lockhart's sage advice to "present the Army as it is, not as it ought to be or would like to be."[5] The challenges of information management in the contemporary world suggest the latter's advice may well be the best way to proceed in the future, but it does not mean that we should cease to appreciate the variegated past of Army-media relations.

Notes

1. Michael S. Sweeney, *The Military and the Press: An Uneasy Truce* (Evanston, IL: Northwestern University Press, 2006), 5.

2. A good example of this view can be found in an article in *Military Review* written shortly before this study went to press. See Lieutenant General William B. Caldwell IV, US Army; Mr. Dennis Murphy; and Mr. Anton Menning, "Learning to Leverage New Media: The Israeli Defense Forces in Recent Conflicts," *Military Review* (May–June 2009): 2–9.

3. Executive Order No. 13283, "Establishing the Office of Global Communications, 21 January 2003," John T. Woolley and Gerhard Peters, *The American Presidency Project* [online] (Santa Barbara, CA: University of California (hosted), Gerhard Peters (database)). http://www.presidency.ucsb.edu/ws/?pid=61379 (accessed 5 May 2009).

4. John F. Marszalek, *Sherman's Other War: The General and the Civil War Press* (Kent, OH: Kent State University Press, 1999), 37–38.

5. Survey of the Operations of the Bureau of Public Relations of the War Department, transmitted to Major General Floyd Park by Jack H. Lockhart, 3 June 1946. Papers of Floyd H. Parks, Box 10, Dwight D. Eisenhower Library.

Glossary

AEF	Army Expeditionary Force
AFHQ	Allied Forces Headquarters
AP	Associated Press
ARVN	Army of the Republic of Vietnam
CENTCOM	US Central Command
CINCPAC	Commander in Chief, Pacific
CPI	Committee on Public Information
DDRS	Declassified Documents Reference System
DOD	Department of Defense
FBI	Federal Bureau of Investigation
FRUS	Foreign Relations of the United States
FUSAG	First United States Army Group
GVN	Government of Vietnam
IPI	International Public Information
IPIG	International Public Information Group
JIB	Joint Information Bureau
MAAG	Military Assistance Advisory Group
MACV	Military Assistance Command, Vietnam
NBC	National Broadcasting Company
NSAM	National Security Action Memorandum
NSC	National Security Council
OASD(PA)	Office of the Assistant Secretary of Defense (Public Affairs)
OPI	Office of Public Information
OSI	Office of Strategic Influence
OWI	Office of War Information
PDD	Presidential Decision Directive
PR	public relations
PRO	Public Relations Officer
ROK	Republic of Korea
SCAP	Supreme Commander Allied Powers
SHAEF	Supreme Headquarters Allied Expeditionary Force
SPG	Special Planning Group
STRATCOM	Strategic Command
UN	United Nations
UP	United Press
US	United States
USIA	United States Information Agency

Bibliography

Primary Sources

Basler, Roy P., ed. "Lincoln-Douglas Debate at Ottawa" (21 August 1958), in *The Collected Works of Abraham Lincoln* (New Brunswick, NJ: Rutgers University Press, 1953).

Chandler, Alfred E., ed. *The Papers of Dwight David Eisenhower.* 5 vols. Baltimore, MD: The Johns Hopkins Press, 1970.

Correspondence Relating to the War with Spain, 2 vols. Washington, DC: Center of Military History, 1993.

Declassified Documents Reference System (DDRS). Woodbridge, CT: Research Publications International, 1992.

Dwight D. Eisenhower Library, Abilene, KS. Papers of Floyd H. Parks.

Eisenhower's Communiqués Digital Collection. http://contentdm.lib.byu.edu/cdm4/document.php?CISOROOT=/EisenhowerCommuniques&CISOPTR=652&REC=2&CISOSHOW=0 (accessed 14 May 2009).

John F. Kennedy Presidential Library & Museum. National Security Act Memorandum (NSAM) 111: First Phase of Viet-NAM Program, 22 November 1961. http://www.jfklibrary.org/Historical+Resources/Archives/Reference+Desk/NSAMs.htm (accessed 14 May 2009).

Link, Arthur S., ed. *The Papers of Woodrow Wilson*, Vols. 41 and 42. Princeton, NJ: Princeton University Press, 1983.

Lyndon Baine Johnson Library and Museum. National Security File. http://www.fas.org/irp/offdocs/nsam-lbj/index.html; and http://www.lbjlib.utexas.edu/johnson/archives.hom/nsams/nsamhom.asp (accessed 1 July 2008).

Marshall Papers. Fort Leavenworth, KS: Combined Arms Research Library. Microfilm, reel 24.

Quadrennial Defense Review Report. Washington, DC: Department of Defense, 2006.

Ronald Reagan Library. http://www.fas.org/irp/offdocs/nsdd/nsdd-077.htm (accessed 1 July 2008).

U.S. Congress. Senate. Committee on Governmental Affairs. "Pentagon Rules on Media Access to the Persian Gulf War." *Hearing before the Committee on Governmental Affairs United States Senate*, 102d Cong., 1st sess., 20 February 1991. Washington, DC: US Government Printing Office, 1991.

US Department of State. *Foreign Relations of the United States [FRUS], 1951. National Security Affairs; Foreign Economic Policy, Vol. I.* Washington, DC: US Government Printing Office, 1951. http://digital.library.wisc.edu/1711.dl/FRUSA.FRUS1951v01 (accessed 13 May 2009).

———. http://www.fas.org/irp/offdocs/nsam-lbj/index.html (accessed 1 July 2008).

US National Strategy for Public Diplomacy and Strategic Communications. Washington, DC: Strategic Communication and Public Diplomacy Coordinating Committee, 2007.

Memoirs

Butcher, Harry C. *My Three Years with Eisenhower*. New York, NY: Simon and Schuster, 1946.

Clinton, William J. *My Life.* New York, NY: Alfred A. Knopf, 2004.

Collins, J. Lawton. *Lightning Joe: An Autobiography*. Baton Rouge, LA: Louisiana State University Press, 1979.

Creel, George. *Rebel at Large: Recollections of Fifty Crowded Years*. New York, NY: G.P. Putnam's Sons, 1947.

Eisenhower, Dwight David. *At Ease: Stories I Tell to Friends*. Garden City, NY: Doubleday & Company, 1967.

Murphy, Robert. *Diplomat Among Warriors*. Garden City, NY: Doubleday & Company, 1964.

Oldfield, Barney. *Never a Shot in Anger*. New York, NY: Duell, Sloan and Pearce, 1956.

Palmer, Frederick. *With My Own Eyes: A Personal Story of Battle Years*. Indianapolis, IN: Bobbs-Merrill, 1933.

Pershing, John J. *My Experiences in the First World War*. New York, NY: Da Capo Press, 1995.

Powell, Colin, with Joseph E. Persico. *My American Journey.* New York, NY: Random House, 1995.

Quirk, Rory. *Wars and Peace: The Memoirs of an American Family*. Novato, CA: Presidio Press, 1999.

Newspapers and Journals

Army Information Digest
Boston Daily Globe
Chicago Daily Tribune
Christian Science Monitor
Los Angeles Times
Military Review
New York Times
Washington Post

Secondary Sources

An Act for Establishing Rules and Articles for the Government of the Armies of the United States, Article 57. http://freepages.military.rootsweb.ancestry.com/~pa91/cfawar.html (accessed 5 May 2009).

Anderson, David L., and John Ernst, eds. *The War That Never Ends: New Perspectives on the Vietnam War*. Lexington, KY: University Press of Kentucky, 2007.

Appleman, Roy E. *Ridgway Duels for Korea.* College Station, TX: Texas A&M University Press, 1990.

Balfour, Michael. *Propaganda in War 1939–1945: Organisations, Policies and Publics in Britain and Germany*. London: Routledge & Kegan Paul, 1979.

Bernhard, Nancy E. "Clearer than Truth: Public Affairs Television and the State Department's Domestic Information Campaigns, 1947–1952." *Diplomatic History* 21, No. 4 (Fall 1997): 545–567.

———. *U.S. Television News and Cold War Propaganda, 1947–1960*. New York, NY: Cambridge University Press, 1999.

Blumenson, Martin. *Mark Clark: The Last of the Great World War II Commanders*. New York, NY: Congdon & Weed, 1984.

Braestrup, Peter. *Big Story: How the American Press and Television Reported and Interpreted the Crisis of Tet 1968 in Vietnam and Washington*, 2 vols. Boulder, CO: Westview Press, 1977.

Brown, Charles H. *The Correspondents War: Journalists in the Spanish-American War*. New York, NY: Charles Scribner's Sons, 1967.

Browne, Malcolm. *The New Face of War*. Indianapolis, IN: Bobbs-Merrill, 1965.

Campen, Alan D., ed. *The First Information War: The Story of Communications, Computers, and Intelligence Systems in the Persian Gulf War*. Fairfax, VA: AFCEA International Press, 1992.

Casey, Steven. *Cautious Crusade: Franklin D. Roosevelt, American Public Opinion, and the War against Nazi Germany*. Oxford, NY: Oxford University Press, 2001.

———. *Selling the Korean War: Propaganda, Politics, and Public Opinion in the United States, 1950–1953*. Oxford, NY: Oxford University Press, 2008.

Chambers, John W. II., ed. *The Oxford Companion to American Military History*. Oxford, NY: Oxford University Press, 1999.

Coats, Stephen D. *Gathering at the Golden Gate: Mobilizing for War in the Philippines*. Fort Leavenworth, KS: Combat Studies Institute Press, 2006.

Cole, Ronald H. *Operation Urgent Fury: The Planning and Execution of Joint Operations in Grenada, 12 October–2 November 1983*. Washington, DC: Joint History Office, Office of the Chairman of the Joint Chiefs of Staff, 1997.

Collins, J. Lawton. "An Information Policy for the New Army." *Army Information Digest* 1, No. 1 (May 1946): 3–5.

Condit, Doris M. *History of the Office of the Secretary of Defense, Vol. II, The Test of War, 1950–1953*. Washington, DC: Historical Office, Office of the Secretary of Defense, 1988.

Corke, Sarah-Jane. *US Covert Operations and Cold War Strategy: Truman, Secret Warfare, and the CIA,1945–53*. London: Routledge, 2008.

Cornell University Law School. http://www4.law.cornell.edu/uscode/22/1461.html (accessed 12 May 2009).

Cosmas, Graham A. *An Army for Empire: The United States Army in the Spanish-American War*. Columbia, MO: University of Missouri Press, 1971.

Coughlin, William J. *Conquered Press: The MacArthur Era in Japanese Journalism*. Palo Alto, CA: Pacific Books, 1952.

Dallek, Robert. *Franklin D. Roosevelt and American Foreign Policy, 1932–1945*. New York, NY: Oxford University Press, 1979.

Darley, William M. "The Missing Component of U.S. Strategic Communications." *Joint Force Quarterly* 47 (Fourth Quarter, 2007): 109–113.

Davis, Robert T. II. "Cold War Interagency Relations and the Struggle to Coordinate Psychological Strategy." In *The US Army and the Interagency Process: Historical Perspectives*, edited by Kendall D. Gott and Michael G. Brooks, 301–319. Fort Leavenworth, KS: Combat Studies Institute Press, 2009.

Denton, Robert E. Jr. *The Media and the Persian Gulf War*. Wesport, CT: Praeger, 1993.

———. http://www.state.gov/www/about_state/history/vol_ii_1961-63/g.html (accessed 12 May 2009).

Desmond, Robert W. *Tides of War: World News Reporting, 1940–1945*. Iowa City, IA: University of Iowa Press, 1984.

D'Este, Carlo. *Eisenhower: A Soldier's Life*. New York, NY: Henry Holt and Company, 2002.

Dower, John W. *War Without Mercy: Race and Power in the Pacific War*. New York, NY: Pantheon Books, 1986.

Eisenhower, Dwight D. *Crusade in Europe*. Garden City, NY: Doubleday & Company, 1948.

Feith, Douglas J. *War and Decision: Inside the Pentagon at the Dawn of the War on Terrorism.* New York, NY: Harper, 2008.

Fialka, John J. *Hotel Warriors: Covering the Gulf War*. Washington, DC: Woodrow Wilson Center Press, 1991.

Fielding, Raymond. *The American Newsreel, 1911–1967*. Norman, OK: University of Oklahoma Press, 1972.

Flint, Roy K. "Task Force Smith and the 24th Division: Delay and Withdrawal, 5–19 July 1950." In *America's First Battles*, edited by Charles E. Heller and William A. Stofft. Lawrence, KS: University Press of Kansas, 1986.

Glen, John. "Journalistic Impedimenta: William Tecumseh Sherman and Free Expression." In *The Civil War and the Press*, edited by David Sachsman, S. Kittrell Rushing, and Debra Reddin van Tuyll, 408–408. New Brunswick, NJ: Transaction Publishers, 2000.

Grose, Peter. *Operation Rollback: America's Secret War behind the Iron Curtain.* Boston, MA: Houghton Mifflin, 2000.

Gunrow, Carl D. "Winning the Information War." *Army* (April 2007): 12–15.

Halberstam, David. *The Making of a Quagmire*. New York, NY: Random House, 1965.

Hallin, Daniel C. *The "Uncensored War": The Media and Vietnam*. New York, NY: Oxford University Press, 1986.

Halloran, Richard. "Strategic Communication." *Parameters* 37 (Autumn 2007): 4–14.

Hammond, William M. *Reporting Vietnam: Media and Military at War*. Lawrence, KS: University Press of Kansas, 1998.

———. *Public Affairs: The Military and the Media, 1962–1968.* Washington, DC: Center of Military History, US Army, 1988.

Hesketh, Roger. *Fortitude: The D-Day Deception Campaign.* London: St. Ermin's Press, 1999.

Hixson, Walter. *Parting the Curtain: Propaganda, Culture and the Cold War, 1945–1961.* New York, NY: St. Martin's Press, 1997.

Horten, Gerd. *Radio Goes to War: The Cultural Politics of Propaganda during World War II*. Berkeley, CA: University of California Press, 2002.

International Public Information. http://www.fas.org/irp/offdocs/pdd/pdd-68.htm (accessed 12 May 2009).

James, D. Clayton. *The Years of MacArthur*, 3 vols. Boston, MA: Houghton Mifflin, 1970–85.

Jessup, John E., ed. *Encyclopedia of the American Military*, Vol. III. New York, NY: Charles Scribner's Sons, 1994.

Jones, Jeffrey B. "Strategic Communication: A Mandate for the United States." *Joint Force Quarterly* 39 (Fourth Quarter, 2005): 108–114.

Keeton, Pamela, and Mark McCann. "Information Operations, STRATCOM, and Public Affairs." *Military Review* 85, No. 6 (November–December 2005): 83–86.

Kennedy, David M. *Over Here: The First World War and American Society*. New York, NY: Oxford University Press, 1980.

Kinnard, Douglas. *The War Managers.* Hanover, NH: University of New England Press, 1977.

Kirby, Edward M., and Jack W. Harris, *Star-Spangled Radio*. Chicago, IL: Ziff-Davis Publishing, 1948.

Krugler, David F. *The Voice of America and the Domestic Propaganda Battles, 1945–1953*. Columbia, MO: University of Missouri Press, 2000.

Krulak, Charles C. "The United States Marine Corps in the 21st Century." *RUSI Journal* (August 1996): 23–26.

Laurie, Clayton D. *The Propaganda Warriors: America's Crusade against Nazi Germany*. Lawrence, KS: University Press of Kansas, 1996.

Leuchtenburg, William E. "Franklin D. Roosevelt: The First Modern President." In *Leadership in the Modern Presidency*, edited by Fred I. Greenstein. Cambridge, MA: Harvard University Press, 1988.

Linn, Brian McAllister. *The Philippine War, 1899–1902*. Lawrence, KS: University Press of Kansas, 2000.

Lord, Carnes. "On the Nature of Strategic Communications." *Joint Force Quarterly* 46 (Third Quarter, 2007): 87–89.

Lucas, Scott. *Freedom's War: The American Crusade against the Soviet Union.* New York, NY: New York University Press, 1999.

MacArthur, John R. *Second Front: Censorship and Propaganda in the Gulf War*. New York, NY: Hill and Wang, 1992.

Marszalek, John F. *Sherman's Other War: The General and the Civil War Press*. Kent, OH: Kent State University Press, 1999.

Mazaar, Michael J., Jeffrey Shaffer, and Benjamin Ederington. *Military Technical Revolution: A Structural Framework*. Washington, DC: Center for Strategic and International Studies, 1993.

McPherson, James M. *Battle Cry of Freedom: The Civil War Era*. New York, NY: Oxford University Press, 1988.

Metzl, Jamie Frederic. "Popular Diplomacy." *Daedalus* 128, No. 2 (Spring, 1999): 177–192.

Mecklin, John. *Mission in Torment: An Intimate Account of the U.S. Role in Vietnam*. Garden City, NY: Doubleday & Company, 1965.

Mitrovich, Gregory. *Undermining the Kremlin: America's Strategy to Subvert the Soviet Bloc, 1947–1956*. Ithaca, NY: Cornell University Press, 2000.

Morris, Edmund. *Theodore Rex*. New York, NY: Random House, 2001.

Murphy, Dennis M., and James F. White. "Propaganda: Can a Word Decide a War?" *Parameters* 37 (Autumn, 2007): 15–27.

Nelson, Michael. *War of the Black Heavens: The Battles of Western Broadcasting in the Cold War*. Syracuse, NY: Syracuse University Press, 1997.

Ninkovich, Frank. *U.S. Information Policy and Cultural Diplomacy*. Washington, DC: Foreign Policy Association, 1996.

Osgood, Kenneth. *Total Cold War: Eisenhower's Secret Propaganda Battle at Home and Abroad*. Lawrence, KS: University Press of Kansas, 2006.

Paddock, Alfred H. Jr. *U.S. Army Special Warfare: Its Origins*, rev. ed. Lawrence, KS: University Press of Kansas, 2002.

Paludan, Phillip S. "'The Better Angels of Our Nature': Lincoln, Propaganda, and Public Opinion in the North during the Civil War." In *On the Road to Total War: The American Civil War and the German Wars of Unification, 1861–1871*, edited by Stig Förster and Jörg Nagler, 357–376. Cambridge, MA: Cambridge University Press, 1997.

Parry-Giles, Shawn J. *The Rhetorical Presidency, Propaganda and the Cold War, 1945–1955*. Westport, CT: Praeger, 2002.

Perret, Geoffrey. *Old Soldiers Never Die: The Life of Douglas MacArthur*. New York, NY: Random House, 1996.

Pogue, Forrest C. *George C. Marshall: Organizer of Victory*. New York, NY: The Viking Press, 1973.

———. *The Supreme Command*. Washington, DC: Center of Military History, 1989 reprint.

Pulwers, Jack E. *The Press of Battle: The GI Reporters and the American People*. Raleigh, NC: Ivy House Publishing, 2003.

Reed, Luther J. "Tokyo Reporting." *Army Information Digest* 1, No. 4 (August 1946): 17–19.

Roeder, George H. Jr. *The Censored War: American Visual Experience during World War II*. New Haven, CT: Yale University Press, 1993.

Ross, Stewart Halsey. *Propaganda for War: How the United States Was Conditioned to Fight the Great War of 1914–1918*. Jefferson, NC: McFarland & Company, 1996.

Schmitz, David F. *The Tet Offensive: Politics, War, and Public Opinion*. Lanham, MD: Rowman & Littlefield, 2005.

Schulzinger, Robert D. *A Time for War: The United States and Vietnam, 1941–1975*. New York, NY: Oxford University Press, 1997.

Smith, Jeffrey A. *War and Press Freedom: The Problem of Prerogative Power.* New York, NY: Oxford University Press, 1999.

Sorensen, Thomas C. *The Word War: The Story of American Propaganda*. New York, NY: Harper & Row, 1968.

Stech, Frank J. "Winning CNN Wars." *Parameters* (Autumn, 1994): 37–56.

Steele, Richard W. "Preparing the Public for War: Efforts to Establish a National Propaganda Agency, 1940–41." *The American Historical Review* 75, No. 6 (October 1970).

Sweeney, Michael S. *The Military and the Press: An Uneasy Truce*. Evanston, IL: Northwestern University Press, 2006.

———. *Secrets of Victory: The Office of Censorship and the American Press and Radio in World War II*. Chapel Hill, NC: University of North Carolina Press, 2001.

Taylor, Philip M. *Munitions of the Mind: A History of Propaganda from the Ancient World to the Present Day*. Manchester, UK: Manchester University Press, 2003.

"The Army Information School Makes a Bow." *Army Information Digest* 1, No. 1 (May 1946): 14–19.

Thomas, Benjamin P., and Harold M. Hyman. *Stanton: The Life and Times of Lincoln's Secretary of War.* New York, NY: Alfred A. Knopf, 1962.

Traxel, David. *Crusader Nation: The United States in Peace and the Great War, 1898–1920.* New York, NY: Alfred A. Knopf, 2006.

Twentieth Century Fund. *Battle Lines: Report of the Twentieth Century Fund on the Military and the Media*. New York, NY: Priority Press, 1985.

Vandiver, Frank E. *Black Jack: The Life and Times of John J. Pershing, Vol. II*. College Station, TX: Texas A&M University Press, 1977.

Vaughn, Stephen. *Holding Fast the Inner Lines: Democracy, Nationalism, and the Committee on Public Information*. Chapel Hill, NC: University of North Carolina Press, 1980.

Wass de Czege, Huba. "On Winning Hearts and Minds." *Army* (August 2006): 8–18.

Watson, Mark Skinner. *Chief of Staff: Prewar Plans and Preparations* [United States Army in World War II, The War Department]. Washington, DC: Center of Military History, 1991.

Weigley, Russell F. *History of the United States Army*, enlarged ed. Bloomington, IN: University of Indiana Press, 1984.

Weintraub, Stanley. *MacArthur's War: Korea and the Undoing of an American Hero*. New York, NY: The Free Press, 2000.

Wilbanks, James H. *The Tet Offensive: A Concise History*. New York, NY: Columbia University Press, 2007.

Wilkerson, Marcus M. *Public Opinion and the Spanish-American War: A Study in War Propaganda*. New York, NY: Russell & Russell, 1932.

Winfield, Betty Houchin. *FDR and the News Media*. Urbana, IL: University of Illinois Press, 1990.

Winkler, Allan M. *The Politics of Propaganda: The Office of War Information, 1942–1945*. New Haven, CT: Yale University Press, 1978.

Woolley, John T., and Gerhard Peters. *The American Presidency Project* [online]. Santa Barbara, CA: University of California (hosted), Gerhard Peters (database). http://www.presidency.ucsb.edu/ws/?pid=69797 (accessed 5 May 2009).

Wooster, Robert. *Nelson A. Miles and the Twilight of a Frontier Army*. Lincoln, NE: University of Nebraska Press, 1993.

Wyatt, Clarence R. "The Media and the Vietnam War." In *The War That Never Ends: New Perspectives on the Vietnam War*, edited by David L. Anderson and John Ernst. Lexington, KY: University Press of Kentucky, 2007.

Young, Peter R., ed. *Defence and the Media in Time of Limited War*. London: Frank Cass, 1992.

Appendix A

Documents Relating to US Government Information Policy

Civil War

Executive Order, August 7, 1861

Executive Order—Taking into Military Possession all Telegraph Lines in the United States, February 25, 1862

World War I

Executive Order 2585—Taking Over Necessary and Closing Unnecessary Radio Stations, April 6, 1917

Executive Order 2594—Creating Committee on Public Information, April 13, 1917

Executive Order 2604—Censorship of Submarine Cables, Telegraph and Telephone Lines, April 28, 1917

World War II

Executive Order 9182—Establishing the Office of War Information, June 13, 1942

Dwight D. Eisenhower, 8 May 1944

Post-World War II

Executive Order 9608—Providing for the Termination of the Office of War Information, and for the Disposition of Its Functions and of Certain Functions of the Office of Inter-American Affairs, August 31, 1945

Extract from the Smith-Mundt Act, 1948

Korean War

Harry Truman, Presidential Memorandum, 5 December 1950

Harry Truman, Directive Creating Psychological Strategy Board, April 4, 1951

Vietnam War

State Department Cable 1006, 21 February 1962

National Security Action Memorandum No. 308, June 22, 1964

National Security Action Memorandum No. 313, July 31, 1964

Post-Vietnam

Extract from National Security Action Memorandum 328, 6 April 1965

National Security Decision Directive Number 77, Management of Public Diplomacy Relative to National Security, January 14, 1983

Guidelines for News Media, 14 January 1991

Presidential Decision Directive PDD 68, 30 April 1999

Executive Order 13283—Establishing the Office of Global Communications, January 21, 2003

Abraham Lincoln
XVI *President of the United States: 1861–1865*

Executive Order[1]
August 7, 1861

By the fifty-seventh article of the act of Congress entitled "An act for establishing rules and articles for the government of the armies of the United States," approved April 10, 1806, holding correspondence with or giving intelligence to the enemy, either directly or indirectly, is made punishable by death, or such other punishment as shall be ordered by the sentence of a court-martial. Public safety requires strict enforcement of this article.

It is therefore ordered, That all correspondence and communication, verbally or by writing, printing, or telegraphing, respecting operations of the Army or military movements on land or water, or respecting the troops, camps, arsenals, intrenchments, or military affairs within the several military districts, by which intelligence shall be, directly or indirectly, given to the enemy, without the authority and sanction of the major-general in command, be, and the same are absolutely prohibited, and from and after the date of this order persons violating the same will be proceeded against under the fifty-seventh article of war.

SIMON CAMERON.

Approved:

A. LINCOLN.

---

1. John T. Woolley and Gerhard Peters, *The American Presidency Project* [online] (Santa Barbara, CA: University of California (hosted), Gerhard Peters (database)). http://www.presidency.ucsb.edu/ws/?pid=70012 (accessed 5 May 2009).

Abraham Lincoln
XVI *President of the United States: 1861–1865*

Executive Order—Taking into Military Possession all Telegraph Lines in the United States[2]
February 25, 1862

WAR DEPARTMENT

Ordered, first. On and after the 26th day of February instant the President, by virtue of the act of Congress, takes military possession of all the telegraph lines in the United States.

Second. All telegraphic communications in regard to military operations not expressly authorized by the War Department, the General Commanding, or the generals commanding armies in the field, in the several departments, are absolutely forbidden.

Third. All newspapers publishing military news, however obtained and by whatever medium received, not authorized by the official authority mentioned in the preceding paragraph will be excluded thereafter from receiving information by telegraph or from transmitting their papers by railroad.

Fourth. Edward S. Sanford is made military supervisor of telegraphic messages throughout the United States. Anson Stager is made military superintendent of all telegraph lines and offices in the United States.

Fifth. This possession and control of the telegraph lines is not intended to interfere in any respect with the ordinary affairs of the companies or with private business.

By order of the President:

EDWIN M. STANTON,

Secretary of War.

---

2. John T. Woolley and Gerhard Peters, *The American Presidency Project* [online] (Santa Barbara, CA: University of California (hosted), Gerhard Peters (database)). http://www.presidency.ucsb.edu/ws/?pid=69797 (accessed 5 May 2009).

Woodrow Wilson
XXVIII *President of the United States: 1913–1921*

Executive Order 2585—Taking Over Necessary and Closing Unnecessary Radio Stations[3]

April 6, 1917

Whereas, the Senate and House of Representatives of the United States of America, in Congress assembled, have declared that a state of war exists between the United States and the Imperial German Government; and

Whereas it is necessary to operate certain radio stations for radio communication by the Government and to close other radio stations not so operated, to insure the proper conduct of the war against the Imperial German Government and the successful termination thereof

Now, therefore, it is ordered by virtue of authority vested in me by the Act to Regulate Radio Communication, approved August 13, 1912, that such radio stations within the jurisdiction of the United States as are required for naval communications shall be taken over by the Government of the United States and used and controlled by it, to the exclusion of any other control or use; and furthermore that all radio stations not necessary to the Government of the United States for naval communications, may be closed for radio communication.

The enforcement of this order is hereby delegated to the Secretary of the Navy, who is authorized and directed to take such action in the premises as to him may appear necessary.

This order shall take effect from and after this date.

WOODROW WILSON
THE WHITE HOUSE,
April 6, 1917.

---

3. John T. Woolley and Gerhard Peters, *The American Presidency Project* [online] (Santa Barbara, CA: University of California (hosted), Gerhard Peters (database)). http://www.presidency.ucsb.edu/ws/?pid=75407) (accessed 5 May 2009).

WOODROW WILSON
XXVIII *President of the United States: 1913–1921*

Executive Order 2594—Creating Committee on Public Information[4]

April 13, 1917

I hereby create a Committee on Public Information, to be composed of the Secretary of State, the Secretary of War, the Secretary of the Navy, and a civilian who shall be charged with the executive direction of the Committee.

As Civilian Chairman of this Committee, I appoint Mr. George Creel. The Secretary of State, the Secretary of War, and the Secretary of the Navy are authorized each to detail an officer or officers to the work of the Committee.

WOODROW WILSON
THE WHITE HOUSE,
April 13, 1917.

4. John T. Woolley and Gerhard Peters, *The American Presidency Project* [online] (Santa Barbara, CA: University of California (hosted), Gerhard Peters (database)). http://www.presidency.ucsb.edu/ws/?pid=75409 (accessed 5 May 2009).

Woodrow Wilson
XXVIII *President of the United States: 1913–1921*

Executive Order 2604—Censorship of Submarine Cables, Telegraph and Telephone Lines[5]

April 28, 1917

Whereas, the existence of a state of war between the United States and the Imperial German Government makes it essential to the public safety that no communication of a character which would aid the enemy or its allies shall be had,

Therefore, by virtue of the power vested in me under the Constitution and by the Joint Resolution passed by Congress on April 6, 1917, declaring the existence of a state of war, it is ordered that all companies or other persons, owning, controlling or operating telegraph and telephone lines or submarine cables, are hereby prohibited from transmitting messages to points without the United States, and from delivering messages received from such points, except those permitted under rules and regulations to be established by the Secretary of War for telegraph and telephone lines, and by the Secretary of the Navy for submarine cables,

To these Departments, respectively, is delegated the duty of preparing and enforcing rules and regulations under this order to accomplish the purpose mentioned.

This order shall take effect from this date.

WOODROW WILSON
THE WHITE HOUSE,
April 28, 1917.

---

5. John T. Woolley and Gerhard Peters, *The American Presidency Project* [online]. (Santa Barbara, CA: University of California (hosted), Gerhard Peters (database)). http://www.presidency.ucsb.edu/ws/?pid=75413 (accessed 5 May 2009).

Franklin D. Roosevelt

XXXII *President of the United States: 1933–1945*

Executive Order 9182—Establishing the Office of War Information[6]

June 13, 1942

In recognition of the right of the American people and of all other peoples opposing the Axis aggressors to be truthfully informed about the common war effort, and by virtue of the authority vested in me by the Constitution, by the First War Powers Act, 1941, and as President of the United States and Commander in Chief of the Army and Navy, it is hereby ordered as follows:

1. The following agencies, powers, and duties are transferred and consolidated into an Office of War Information which is hereby established within the Office for Emergency Management in the Executive Office of the President:

(a) The Office of Facts and Figures and its powers and duties.

(b) The Office of Government Reports and its powers and duties.

(c) The powers and duties of the Coordinator of Information relating to the gathering of public information and its dissemination abroad, including, but not limited to, all powers and duties now assigned to the Foreign Information Service, Outpost, Publications, and Pictorial Branches of the Coordinator of Information.

(d) The powers and duties of the Division of Information of the Office for Emergency Management relating to the dissemination of general public information on the war effort, except as provided in paragraph 10.

2. At the head of the Office of War Information shall be a Director appointed by the President. The Director shall discharge and perform his functions and duties under the direction and supervision of the President. The Director may exercise his powers, authorities, and duties through such officials or agencies and in such manner as he may determine.

3. There is established within the Office of War Information a Committee on War Information Policy consisting of the Director as Chairman,

---

6. John T. Woolley and Gerhard Peters, *The American Presidency Project* [online] (Santa Barbara, CA: University of California (hosted), Gerhard Peters (database)). http://www.presidency.ucsb.edu/ws/?pid=16273 (accessed 5 May 2009).

representatives of the Secretary of State, the Secretary of War, the Secretary of the Navy, the Joint Psychological Warfare Committee, and of the Coordinator of Inter-American Affairs, and such other members as the Director, with the approval of the President, may determine. The Committee on War Information Policy shall formulate basic policies and plans on war information, and shall advise with respect to the development of coordinated war information programs.

4. Consistent with the war information policies of the President and with the foreign policy of the United States, and after consultation with the Committee on War Information Policy, the Director shall perform the following functions and duties:

(a) Formulate and carry out, through the use of press, radio, motion picture, and other facilities, information programs designed to facilitate the development of an informed and intelligent understanding, at home and abroad, of the status and progress of the war effort and of the war policies, activities, and aims of the Government.

(b) Coordinate the war informational activities of all Federal departments and agencies for the purpose of assuring an accurate and consistent flow of war information to the public and the world at large.

(c) Obtain, study, and analyze information concerning the war effort and advise the agencies concerned with the dissemination of such information as to the most appropriate and effective means of keeping the public adequately and accurately informed.

(d) Review, clear, and approve all proposed radio and motion picture programs sponsored by Federal departments and agencies; and serve as the central point of clearance and contact for the radio broadcasting and motion-picture industries, respectively, in their relationships with Federal departments and agencies concerning such Government programs.

(e) Maintain liaison with the information agencies of the United Nations for the purpose of relating the Government's informational programs and facilities to those of such Nations.

(f) Perform such other functions and duties relating to war information as the President may from time to time determine.

5. The Director is authorized to issue such directives concerning war information as he may deem necessary or appropriate to carry out the purposes of this Order, and such directives shall be binding upon the several Federal departments and agencies. He may establish by regulation the types and classes of informational programs and releases which shall require clearance and approval by his office prior to dissemination. The Director may

require the curtailment or elimination of any Federal information service, program, or release which he deems to be wasteful or not directly related to the prosecution of the war effort.

6. The authority, functions, and duties of the Director shall not extend to the Western Hemisphere exclusive of the United States and Canada.

7. The formulation and carrying out of informational programs relating exclusively to the authorized activities of the several departments and agencies of the Government shall remain with such departments and agencies, but such informational programs shall conform to the policies formulated or approved by the Office of War Information. The several departments and agencies of the Government shall make available to the Director, upon his request, such information and data as may be necessary to the performance of his functions and duties.

8. The Director of the Office of War Information and the Director of Censorship shall collaborate in the performance of their respective functions for the purpose of facilitating the prompt and full dissemination of all available information which will not give aid to the enemy.

9. The Director of the Office of War Information and the Defense Communications Board shall collaborate in the performance of their respective functions for the purpose of facilitating the broadcast of war information to the peoples abroad.

10. The functions of the Division of Information of the Office for Emergency Management with respect to the provision of press and publication services relating to the specific activities of the constituent agencies of the Office for Emergency Management are transferred to those constituent agencies respectively, and the Division of Information is accordingly abolished.

11. Within the limits of such funds as may be made available to the Office of War Information, the Director may employ necessary personnel and make provision for the necessary supplies, facilities, and services. He may provide for the internal management and organization of the Office of War Information in such manner as he may determine.

12. All records, contracts, and property (including office equipment) of the several agencies and all records, contracts, and property used primarily in the administration of any powers and duties transferred or consolidated by this Order, and all personnel used in the administration of such agencies, powers, and duties (including officers whose chief duties relate to such administration) are transferred to the Office of War Information, for use in the administration of the agencies, powers, and duties transferred

or consolidated by this Order; provided, that any personnel transferred to the Office of War Information by this Order, found by the Director of the Office of War Information to be in excess of the personnel necessary for the administration of the powers and duties transferred to the Office of War Information, shall be retransferred under existing procedure to other positions in the Government service, or separated from the service.

13. So much of the unexpended balances of appropriations, allocations, or other funds available for the use of any agency in the exercise of any power or duty transferred or consolidated by this Order or for the use of the head of any agency in the exercise of any power or duty so transferred or consolidated, as the Director of the Bureau of the Budget with the approval of the President shall determine, shall be transferred to the Office of War Information, for use in connection with the exercise of powers or duties so transferred or consolidated. In determining the amount to be transferred, the Director of the Bureau of the Budget may include an amount to provide for the liquidation of obligations incurred against such appropriations, allocations, or other funds prior to the transfer or consolidation.

Dwight D. Eisenhower, 8 May 1944[7]

TO ALL UNIT COMMANDERS, A.E.F.

Confidential

To all Unit Commanders, Allied Expeditionary Force: At my first Press Conference as Supreme Commander I told the War Correspondents that once they were accredited to my headquarters I considered them quasi-staff officers.

All war correspondents that may accompany the expedition are first accredited to Supreme Headquarters and operate under policies approved by the Supreme Commander. They are, in turn, assigned to lower headquarters in accordance with agreements between the Public Relations Division of this headquarters and the Public Relations Officers on the staffs of the several Commanders-in-Chief. This allocation is always limited by accommodations available. Public Relations Officers of the various echelons act as their guides. As a matter of policy accredited war correspondents should be accorded the greatest possible latitude in gathering legitimate news.

Consequently it is desired that, subject always to the requirements of operations, of which the Commander on the spot must be the sole judge, Commanders of all echelons and Public Relations Officers and Conducting Officers give accredited war correspondents all reasonable assistance. They should be allowed to talk freely with officers and enlisted personnel and to see the machinery of war in operation in order to visualize and transmit to the public the conditions under which the men from their countries are waging war against the enemy.

---

7. Alfred E. Chandler, ed., *The Papers of Dwight David Eisenhower, The War Years, Vol. III,* (Baltimore, MD: The Johns Hopkins Press, 1970), 1853.

Harry S. Truman
XXXIII *President of the United States: 1945–1953*

Executive Order 9608—Providing for the Termination of the Office of War Information, and for the Disposition of Its Functions and of Certain Functions of the Office of Inter-American Affairs[8]

August 31, 1945

By virtue of the authority vested in me by the Constitution and Statutes, including Title I of the First War Powers Act, 1941, and as President of the United States, it is hereby ordered as follows:

1. Effective as of the date of this order:

(a) There are transferred to and consolidated in an Interim International Information Service, which is hereby established in the Department of State, those functions of the Office of War Information (established by Executive Order No. 9182 of June 13, 1942), and those informational functions of the Office of Inter-American Affairs (established as the Office of the Coordinator of Inter-American Affairs by Executive Order No. 8840 of July 30, 1941 and renamed as the Office of Inter-American Affairs by Executive Order No. 9532 of March 23, 1945), which are performed abroad or which consist of or are concerned with informing the people of other nations about any matter in which the United States has an interest, together with so much of the personnel, records, property, and appropriation balances of the Office of War Information and the Office of Inter-American Affairs as the Director of the Bureau of the Budget shall determine to relate primarily to the functions so transferred. Pending the abolition of the said Service under paragraph 3(a) of this order, (1) the head of the Service, who shall be designated by the Secretary of State, shall be responsible to the Secretary of State or to such other officer of the Department as the Secretary shall direct, (2) the Service shall, except as otherwise provided in this order, be administered as an organizational entity in the Department of State, (3) the Secretary may transfer from the Service, to such agencies of the Department of State as he shall designate or establish, any function of the Service, and (4) the Secretary may terminate any function of the

---

8. John T. Woolley and Gerhard Peters, *The American Presidency Project* [online] (Santa Barbara, CA: University of California (hosted), Gerhard Peters (database)). http://www.presidency.ucsb.edu/ws/?pid=60671 (accessed 5 May 2009).

Service, in which event he shall provide for the winding up of the affairs relating to any function so terminated.

(b) There are transferred to the Bureau of the Budget the functions of the Bureau of Special Services of the Office of War Information and functions of the Office of War Information with respect to the review of publications of Federal agencies, together with so much of the personnel, records, and property, and appropriation balances of the Office of War Information as the Director of the Bureau of the Budget shall determine to relate primarily to the said functions.

(c) All those provisions of prior Executive orders which are in conflict with this order are amended accordingly. Paragraph 6 of the said Executive Order No. 8840 and paragraphs 3, 6, and 8 of the said Executive Order No. 9182 are revoked.

2. Effective as of the close of business September 15, 1945:

(a) There are abolished the functions of the Office of War Information then remaining.

(b) The Director of the Office of War Information shall, pending the abolition of the Office of War Information under paragraph 3(b) of this order, proceed to wind up the affairs of the Office relating to such abolished functions.

3. Effective as of the close of business December 31, 1945:

(a) The Interim International Information Service, provided for in paragraph 1(a) of this order, together with any functions then remaining under the Service, is abolished.

(b) The Office of War Information, including the office of the Director of the Office of War Information, is abolished.

(c) There are transferred to the Department of the Treasury all of the personnel, records, property, and appropriation balances of the Interim International Information Service and of the Office of War Information then remaining, for final liquidation, and so much thereof as the Director of the Bureau of the Budget shall determine to be necessary shall be utilized by the Secretary of the Treasury in winding up all of the affairs of the Service.

HARRY S. TRUMAN
THE WHITE HOUSE,
August 31, 1945.

Extract from the Smith-Mundt Act, 1948[9]

Title 22, Section 1461, United States Code, Chapter 18, Subchapter V: Dissemination Abroad of Information about the United States

§ 1461. General authorization.

(a) Dissemination of Information Abroad

The Secretary [of State] is authorized, when he finds it appropriate, to provide for the preparation, and dissemination abroad, of information about the United States, its people, and its policies, through press, publications, radio, motion pictures, and other information media, and through information centers and instructors abroad. Subject to subsection (b) of this section, any such information (other than "Problems of Communism" and the "English Teaching Forum" which may be sold by the Government Printing Office) shall not be disseminated within the United States, its territories, or possessions, but, on request, shall be available in the English language at the Department of State, at all reasonable times following its release as information abroad, for examination only by representatives of United States press associations, newspapers, magazines, radio systems, and stations, and by research students and scholars, and, on request, shall be available for examination only to Members of Congress.

---

9. Cornell University Law School. http://www4.law.cornell.edu/uscode/22/1461.html (accessed 12 May 2009).

Harry Truman, Presidential Memorandum, 5 December 1950[10]

THE WHITE HOUSE
WASHINGTON

December 5, 1950

CONFIDENTIAL

MEMORANDUM FOR:

The Secretary of State
The Secretary of the Treasury
The Secretary of Defense
The Attorney General
The Postmaster General
The Secretary of the Interior
The Secretary of Agriculture
The Secretary of Commerce
The Secretary of Labor
Chairman, National Security Resources Board
Administrator, Economic Cooperation Administration
Director, Central Intelligence Agency
Administrator, Economic Stabilization Agency
Director, Selective Service System

In the light of the present critical international situation, and until further written notice from me, I wish that each one of you would take immediate steps to reduce the number of public speeches pertaining to foreign or military policy made by officials of the departments and agencies of the Executive Branch. This applies to officials in the field as well as those in Washington.

No speech, press release, or other public statement concerning foreign policy should be released until it has received clearance from the Department of State.

No speech, press release, or other public statement concerning military policy should be released until it has received clearance from the Department of Defense.

---

10. Declassified Documents Reference System (DDRS) (Woodbridge, CT: Research Publications International, 1992). Microfiche #127, #1873.

In addition to the copies submitted to the Department of State or Defense for clearance, advance copies of speeches and press releases concerning foreign policy or military policy should be submitted to the White House for information.

The purpose of this memorandum is not to curtail the flow of information to the American people, but rather to insure that the information made public is accurate and fully in accord with the policies of the United States Government.

/s/ HARRY TRUMAN

Harry Truman, Directive Creating Psychological Strategy Board[11]

Directive by the President to the Secretary of State, the Secretary of Defense, and the Director of Central Intelligence:

SECRET WASHINGTON, April 4, 1951

It is the purpose of this directive to authorize and provide for the more effective planning, coordination and conduct, within the framework of approved national policies, of psychological operations.

There is hereby established a Psychological Strategy Board responsible, within the purposes and terms of this directive, for the formulation and promulgation, as guidance to the departments and agencies responsible for psychological operations, of over-all national psychological objectives, policies, programs, and for the coordination and evaluation of the national psychological effort.

The Board will report to the National Security Council on the Board's activities and on its evaluation of the national psychological operations, including implementation of approved objectives, policies, and programs by the departments and agencies concerned.

For the purposes of this directive, psychological operations shall include all activities (other than overt types of economic warfare) envisioned under NSC 59/1 and NSC 10/2, the operational planning and execution of which shall remain, subject to this directive, as therein assigned.

The Board shall be composed of:

a. The Undersecretary of State, the Deputy Secretary of Defense, and the Director of Central Intelligence, or, in their absence, their appropriate designees;

b. An appropriate representative of the head of each such other department or agency of the Government as may, from time to time, be determined by the Board.

The Board shall designate one of its members as Chairman.

A representative of the Joint Chiefs of Staff shall sit with the Board as its principal military adviser in order that the Board may ensure that its

---

11. United States Department of State, *Foreign Relations of the United States, 1951. National Security Affairs: Foreign Economic Policy, Vol. I* (Washington, DC: US Government Printing Office, 1951). http://digital.library.wisc.edu/1711.dl/FRUS.FRUS1951v01 (accessed 13 May 2009).

objectives, policies and programs shall be related to approved plans for military operations.

There is established under the Board a Director who shall be designated by the President and who shall receive compensation of $16,000 per year. The Director shall direct the activities under the Board. In carrying out this responsibility, he shall

a. Be responsible for having prepared the programs, policies, reports, and recommendations for the Board's consideration,

b. Sit with the Board and be responsible to it for organizing its business and for expediting the reaching of decisions,

c. Promulgate the decisions of the Board.

d. Ascertain the manner in which agreed upon objectives, policies, and programs of the Board are being implemented and coordinated among the departments and agencies concerned,

e. Report thereon and on his evaluation of the national psychological operations to the Board together with his recommendations,

f. Perform such other duties necessary to carry out his responsibilities as the Board may direct.

The Director, within the limits of funds and personnel made available by the Board for this purpose, shall organize and direct a staff to assist in carrying out his responsibilities. The Director shall determine the organization and qualifications of the staff, which may include individuals employed for this purpose, including part-time experts, and/or individuals detailed from the participating departments and agencies for assignment to full-time duty or on an *ad hoc* task force basis. Personnel detailed for assignment to duty under the terms of this directive shall be under the control of the Director, subject only to necessary personnel procedures within their respective departments and agencies.

The participating departments and agencies shall afford to the Director and the staff such assistance and access to information as may be specifically requested by the Director in carrying out his assigned duties.

The heads of the departments and agencies shall examine into the present arrangements within their departments and agencies for the conduct, direction and coordination of psychological operations with a view toward readjusting or strengthening them if necessary to carry out the purposes of this directive. The Secretary of State is authorized to effect such readjustments in the organization established under NSC 59/1 as he deems necessary to accomplish the purposes of this directive.

This directive does not authorize the Board nor the Director to perform any "psychological operations."

In performing its functions, the board shall utilize to the maximum extent the facilities and resources of the participating departments and agencies.

/s/ Harry S. Truman

State Department Cable 1006, 21 February 1962[12]

1006. Joint State-Defense-USIA Message. Embtel 1013. State, Defense, USIA concur in view that more flexibility needed at local level in handling of American newsmen covering Viet-Nam operations. We conclude that in absence of rigid censorship, US interests best be protected through policy of maximum feasible cooperation, guidance and appeal to good faith of correspondents.

Recent press and magazine reports are convincing evidence that speculation stories by hostile reporters often more damaging than facts they might report.

Ambassador has over-all authority for handling of newsmen, in so far as US is concerned. He will make decisions as to when newsmen permitted to go on any missions with US personnel, when approved by US military commander. They also must approve in advance transport of newsmen on US ships and other US craft, including air, involved in Viet-Nam operations. Ambassador should coordinate information policy with GVN if possible.

Attention called to the following guidelines which we believe in our national interest. US military and civilian personnel must see that they are adhered to scrupulously and that Ambassador given complete cooperation if we [are] to avoid harmful press repercussions on both domestic and international scene.

1. This is not a US war. US personnel, civilian or military, should not grant interviews or take other actions implying all-out US involvement. Important that we constantly reinforce the idea that this is struggle in which tens of thousands Vietnamese [are] fighting for their freedom, and that our participation is only in training, advisory and support phases.

2. We recognize it natural that American newsmen will concentrate on activities of Americans. It is not in our interest, however, to have stories indicating that Americans are leading and directing combat missions against the Viet Cong.

3. Should impress upon newsmen that purpose of certain classified operations is to flush out and destroy bands of vicious Viet Cong terrorists. Every effort made to avoid harming innocent civilians. Sensational press stories about children or civilians who become unfortunate victims of military operations are clearly inimical to national interest.

---

12. United States Department of State. http://www.state.gov/www/about_state/history/vol_ii_1961-63/g.html (accessed 12 May 2009).

4. Operations may be referred to in general terms, but specific numbers—particularly numbers of Americans involved—and details of material introduced are not to be provided. On tactical security matters, analysis strength and weaknesses and other operational details which might aid enemy should be avoided.

5. We cannot avoid all criticism of Diem. No effort should be made to "forbid" such articles. Believe, however, that if newsmen feel we are cooperating they will be more receptive to explanation that we [are] in a vicious struggle where support of South Vietnamese is crucial and that articles that tear down Diem only make our task more difficult.

6. Emphasize to newsmen fact that success of operation requires high level GVN-American cooperation and that frivolous, thoughtless criticism of GVN makes cooperation difficult [to] achieve.

7. Correspondents should not be taken on missions whose nature [is] such that undesirable dispatches would be highly probable.

Think it advisable that Ambassador and General Harkins see newsmen at frequent intervals so as to establish point that they [are] keeping press informed to extent compatible with security. Should consider pre-operations briefing of newsmen by designated spokesman.

The point below for consideration and private use at Ambassador's discretion:

It should be possible for Ambassador and/or military to exact from responsible correspondents voluntary undertakings to avoid emphasis in dispatches of sensitive matters, to check doubtful facts with US Government authorities on scene. Seriousness of need for this may be duly impressed on responsible correspondents to extent that, in interests of national security and their own professional needs, they can be persuaded to adopt self-policing machinery. Can be reminded that in World War II American press voluntarily accepted broad and effective censorship. In type struggle now going on in Viet-Nam such self-restraint by press no less important. Important to impress on newsmen that at best this is long term struggle in which most important developments may be least sensational in which "decisive battles" are most unlikely, therefore stories implying sensational "combat" each day are misleading.

Additional press guidance will be provided from time to time. Your reactions and suggestions are welcome.

Rusk

THE WHITE HOUSE

WASHINGTON

June 22, 1964

NATIONAL SECURITY ACTION MEMORANDUM NO. 308[13]

MEMORANDUM TO

The Secretary of State
The Secretary of Defense
The Director of Central Intelligence
The Director, U.S. Information Agency
The Administrator, Agency for International Development

Domestic understanding and support are essential to the success of United States operations in Southeast Asia. The national interest requires full understanding of our policy and purpose in this area. I am not satisfied with the performance of the several departments in this area; we require stronger arrangements.

I have therefore designated Mr. Robert J. Manning, Assistant Secretary of State for Public Affairs, to generate and to coordinate a broad program to bring the American people a complete and accurate picture of the United States involvement in Southeast Asia, and to show why this involvement is essential.

Mr. Manning is instructed to draw as necessary upon the resources of all government agencies in obtaining and disseminating the facts needed by the American people. He will call as necessary upon the senior policy and information officers of your agency. I request that you take the steps necessary to ensure that he receives the unstinting cooperation on a priority basis. I have instructed Mr. Manning to inform me if he encounters any delays or obstructions.

/s/ Lyndon B. Johnson

---

13. United States Department of State. http://www.fas.org/irp/offdocs/nsam-lbj/index.html (accessed 1 July 2008); Lyndon Baine Johnson Library and Museum. http://www.lbjlib.utexas.edu/johnson/archives.hom/nsams/nsamhom.asp (accessed 1 July 2008).

THE WHITE HOUSE

WASHINGTON

July 31, 1964

NATIONAL SECURITY ACTION MEMORANDUM NO. 313[14]

MEMORANDUM FOR

THE SECRETARY OF STATE
THE SECRETARY OF DEFENSE
THE DIRECTOR OF CENTRAL INTELLIGENCE

The President has noticed this week a number of stories on Southeast Asia with Washington datelines which give the impression that some members of the Government are giving conflicting and mutually inconsistent documents to reporters, and that there may be some unauthorized use of information drawn from classified cables. The President requests that each Department and Agency head take further appropriate measures to impress upon all personnel with access to reporters that public comment on this subject should be most carefully handled as set forth in NSAM 308.

/s/ McGeorge Bundy

14. Johnson Library, National Security File. http://www.fas.org/irp/offdocs/nsam-lbj/index.html; and http://www.lbjlib.utexas.edu/johnson/archives.hom/nsams/nsamhom.asp (accessed 1 July 2008).

Extract from National Security Action Memorandum 328, 6 April 1965[15]

[Points 1–4 reiterated previous Presidential approval of nonmilitary, psychological, covert, and military programs of action for Vietnam.]

* * * * * * *

5. The President approved an 18-20,000 man increase in U.S. military support forces to fill out existing units and supply needed logistic personnel.

6. The President approved the deployment of two additional Marine Battalions and one Marine Air Squadron and associated headquarters and support elements.

7. The President approved a change of mission for all Marine Battalions deployed to Vietnam to permit their more active use under conditions to be established and approved by the Secretary of Defense in consultation with the Secretary of State.

* * * * * * *

11. The President desires that with respect to the actions in paragraphs 5 through 7, premature publicity be avoided by all possible precautions. The actions themselves should be taken as rapidly as practicable, but in ways that should minimize any appearance of sudden changes in policy, and official statements on these troop movements will be made only with the direct approval of the Secretary of Defense, in consultation with the Secretary of State. The President's desire is that these movements and changes should be understood as being gradual and wholly consistent with existing policy.

/s/ McGeorge Bundy

---

15. Johnson Library, National Security File. http://www.fas.org/irp/offdocs/nsam-lbj/index.html; and http://www.lbjlib.utexas.edu/johnson/archives.hom/nsams/nsamhom.asp (accessed 1 July 2008).

THE WHITE HOUSE

WASHINGTON

January 14, 1983

NATIONAL SECURITY DECISION DIRECTIVE NUMBER 77[16]

MANAGEMENT OF PUBLIC DIPLOMACY RELATIVE TO NATIONAL SECURITY

I have determined that it is necessary to strengthen the organization, planning and coordination of the various aspects of public diplomacy of the United States Government relative to national security. Public diplomacy is comprised of those actions of the U.S. Government designed to generate support of our national security objectives.

A Special Planning Group (SPG) under the National Security Council will be established under the chairmanship of the Assistant to the President for National Security Affairs. Membership shall consist of the Secretary of State, Secretary of Defense, the Director of the United States Information Agency, the Director of the Agency of International Development, and the Assistant to the President for Communications or their designated alternate. Other senior White House officials will attend as appropriate. Senior representatives of other agencies may attend at the invitation of the chairman.

The SPG shall be responsible for the overall planning, direction, coordination and monitoring of implementation of public diplomacy activities. It shall ensure that a wide-ranging program of effective initiatives is developed and implemented to support national security policy, objectives and decisions. Public diplomacy activities involving the President or the White House will continue to be coordinated with the Office of the White House Chief of Staff.

Four interagency standing committees will be established, and report directly to the SPG. The SPG will ensure that guidance to these committees is provided, as required, so that they can carry out their responsibilities in

16. Ronald Reagan Library. http://www.fas.org/irp/offdocs/nsdd/nsdd-077.htm (accessed 1 July 2008).

the area of public diplomacy. The SPG will further periodically review the activities of the four permanent coordinating committees to insure that plans are being implemented and that resource commitments are commensurate with established priorities.

The NSC Staff, in consultation with the regular members of the SPG, will provide staff support to the SPG and facilitate effective planning, coordinating and implementing of plans and programs approved by the SPG. The NSC Staff will call periodic meetings of the four committee chairmen or their designees to ensure inter-committee coordination.

—*Public Affairs Committee*: This coordinating committee will be co-chaired by the Assistant to the President for Communications and the Deputy Assistant to the President for National Security Affairs. This group will be responsible for the planning and coordinating on a regular basis of U.S. Government public affairs activities relative to national security. Specifically, it will be responsible for the planning and coordination of major speeches on national security subjects and other public appearances by senior officials, and for planning and coordination with respect to public affairs matters concerning national security and foreign policy events and issues with foreign and domestic dimensions. This committee will coordinate public affairs efforts to explain and support major U.S. foreign policy initiatives.

—*International Information Committee*: This committee will be chaired by a senior representative of the United States Information Agency. A senior representative of the Department of State shall serve as vice chairman of the committee. The body will be responsible for the planning, coordinating and implementing international information activities in support of U.S. policies and interests relative to national security. It will assume the responsibilities of the existing "Project Truth" Policy Group. The committee shall be empowered to make recommendations and, as appropriate, to direct the concerned agencies, interagency groups and working groups with respect to information strategies in key policy areas, and it will be responsible for coordinating and monitoring implementation of strategies on specific functional or geographic areas.

—*International Political Committee*: This committee will be established under the chairmanship of a senior representative of the Department of State. A senior representative of the United States Information Agency shall serve as vice chairman of the committee. This group will be responsible for planning, coordinating and implementing international political activities in support of United States policies and interests relative to national security. Included among such activities are aid, training and organizational support for foreign governments and private groups to encourage the growth of democratic political institutions and practices. This will require close collaboration with other foreign policy efforts—diplomatic, economic, military—as well as a close relationship with those sectors of the American society—labor, business, universities, philanthropy, political parties, press—that are or could be more engaged in parallel efforts overseas. This group will undertake to build up the U.S. Government capability to promote democracy, as enunciated in the President's speech in London on June 8, 1982. Furthermore, this committee will initiate plans, programs and strategies designed to counter totalitarian ideologies by the Soviet Union or Soviet surrogates. This committee shall be empowered to make recommendations and, as appropriate, to direct the concerned departments and agencies to implement political action strategies in support of key policy objectives. Attention will be directed to generate policy initiatives keyed to coming events. Close coordination with the other committees will be essential.

—*International Broadcasting Committee*: This committee will be chaired by a representative of the Assistant to the President for National Security Affairs. This committee will be responsible for the planning and coordination of international broadcasting activities sponsored by the U.S. Government consistent with existing statutory requirements and the guidance established by NSDD 45. Among its principal responsibilities will be diplomatic and technical planning relative to modernization of U.S. international broadcasting capabilities, the development of anti-jamming strategies and techniques, planning relative to direct radio broadcast by satellite and longer term considerations of the potential for direct T.V. broadcasting.

Each designated committee is authorized to establish, as appropriate, working groups or ad hoc task forces to deal with specific issues or programs.

All agencies should ensure that the necessary resources are made available for the effective operation of the interagency groups here established.

Implementing procedures for these measures will be developed as necessary.

/s/ Ronald Reagan

Guidelines for News Media, 14 January 1991[17]

News media personnel must carry and support any personal and professional gear they take with them, including protective cases for professional equipment, batteries, cables, converters, etc.

Night Operations—Light discipline restrictions will be followed. The only approved light source is a flashlight with a red lens. No visible light source, including flash or television lights will be used when operating with forces at night unless specifically approved by the on-scene commander.

Because of host-nation requirements, you must stay with your public affairs escort while on Saudi bases. At other U.S. tactical or field locations and encampments, a public affairs escort may be required because of security, safety, and mission requirements as determined by the host commander.

Casualty information, because of concern of the notification of the next of kin, is extremely sensitive. By executive directive, next of kin of all military fatalities must be notified in person by a uniformed member of the appropriate service. There have been instances in which the next of kin have first learned of the death or wounding of a loved one through the news media. The problem is particularly difficult for visual media photographs showing a recognizable face, nametag, or other identifying feature should not be used before next of kin have been notified. The anguish that sudden recognition at home can cause far outweighs the news value of the photograph, film, or videotape. News coverage of casualties in medical centers will be in strict compliance with the instructions of doctors and medical officials.

To the extent that individuals in the news media seek access to the U.S. area of operation, the following rule applies: Prior to or upon commencement of hostilities, media pools will be established to provide initial combat coverage of U.S. forces. U.S. news media personnel present in Saudi Arabia will be given the opportunity to join CENTCOM media pools, providing they agree to pool their products. News media personnel who are not members of official CENTCOM media pools will not be permitted into forward areas. Reporters are strongly discouraged from attempting to link up on their own with combat units. U.S. commanders will maintain extremely tight security throughout the operational area and will exclude from the area of operation all unauthorized individuals.

---

17. Senate Committee on Governmental Affairs, *Pentagon Rules on Media Access to the Persian Gulf War*, 102d Congress, 1st sess., 20 February 1991, v–vi.

For news media personnel participating in designated CENTCOM Media Pools:

(1) Upon registering with the JIB, news media should contact their respective pool coordinator for an explanation of pool considerations.

(2) In the event of hostilities, pool products will be subject to review before release to determine if they contain sensitive information about military plans, capabilities, operations, or vulnerabilities (see attached ground rules) that would jeopardize the outcome of an operation or the safety of U.S. or coalition forces. Material will be examined solely for its conformance to the attached ground rules, not for its potential to express criticism or cause embarrassment. The public affairs escort officer on scene will review pool reports, discuss ground rule problems with the reporter, and in the limited circumstances when no agreement can be reached with a reporter about disputed materials, immediately send the disputed materials to JIB Dharan for review by the JIB Director and the appropriate news media representative. If no agreement can be reached, the issue will be immediately forwarded to OASD(PA) for review with the appropriate bureau chief. The ultimate decision on publication will be made by the originating reporter's news organization.

(3) Correspondents may not carry a personal weapon.

Ground Rules, 14 January 1991

The following information should not be reported because its publication or broadcast could jeopardize operations and endanger lives.

(1) For U.S. or coalition units, specific numerical information on troop strength, aircraft, weapons systems, on-hand equipment,

International Public Information (IPI) Presidential Decision Directive PDD 68[18] 30 April 1999

[No text or factsheet for this PDD has been released.]

> *On 30 April 1999 President Clinton issued a secret Presidential Decision Direction—PDD 68—ordering the creation of an International Public Information (IPI) to address problems identified during military missions in Kosovo and Haiti, when no single US agency was empowered to coordinate US efforts to sell its policies and counteract bad press abroad. The IPI system is geared towards prevention and mitigation of crises and operates on a continuous basis. PDD-68 is evidently intended to replace the provisions of NSDD 77 "Management of Public Diplomacy Relative to National Security" issued by President Reagan on 14 February 1983.*
>
> *PDD 68 ordered top officials from the Defense, State, Justice, Commerce and Treasury departments and the Central Intelligence Agency and FBI to establish an IPI Core Group. The IPI Core Group is chaired by the Under Secretary for Public Diplomacy and Public Affairs at the State Department. The IPI Core Group is ordered by the Presidential Directive to "assist [U.S. government] efforts in defeating adversaries." "The intelligence community will play a crucial role . . . for identifying hostile foreign propaganda and deception that targets the U.S.," the Group's charter says. The IPI Core Group will arrange "training exercises at the National Defense University, National Foreign Affairs Training Center, the Service War Colleges" and other institutions.*
>
> *The International Public Information [IPI] System is designed to "influence foreign audiences" in support of US foreign policy and to counteract propaganda by enemies of the United States. The intent is "to enhance U.S. security, bolster America's economic prosperity and to promote democracy abroad," according to the IPI Core Group Charter. The Group's charter states that IPI control over "international military information" is intended*

18. International Public Information. http://www.fas.org/irp/offdocs/pdd/pdd-68.htm (accessed 12 May 2009).

to "influence the emotions, motives, objective reasoning and ultimately the behavior of foreign governments, organizations, groups and individuals." The IPIG will encourage the United Nations and other international organizations to make "effective use of IPI . . . in support of multilateral peacekeeping." According to the IPIG Charter, IPI activities "are overt and address foreign audiences only" while domestic information should be "deconflicted" and "synchronized" to avoid contradictory messages.

Previously, the US Information Agency and the State Department were the primary agencies with responsibility for international public diplomacy. But with the information revolution, all agencies now have the ability to communicate internationally and interact with foreign populations. IPI is a mechanism that has been established to make sure that these various actors are working in a coordinated manner. According to the IPIG Charter, "The objective of IPI is to synchronize the informational objectives, themes and messages that will be projected overseas . . . to prevent and mitigate crises and to influence foreign audiences in ways favorable to the achievement of U.S. foreign policy objectives." The charter insists that information distributed through IPI should be designed not "to mislead foreign audiences" and that information programs "must be truthful."

The new information policy will not be used to influence the American public, which is prohibited by U.S. law. However, since foreign media reports are frequently reflected in American news media, it will be impossible to entirely preclude a backwash of the IPI-generated information into America. The IPIG Charter recognizes this, calling for the US Government domestic public affairs activities to be coordinated with foreign IPI efforts. According to the IPIG Charter, information aimed at domestic audiences should "be coordinated, integrated, deconflicted and synchronized with the [IPI Core Group] to achieve a synergistic effect for [government] strategic information activities."

George W. Bush

Executive Order 13283—Establishing the Office of Global Communications[19]

January 21, 2003

By the authority vested in me as President by the Constitution and the laws of the United States of America, it is hereby ordered as follows:

Section 1. Establishment of the Office of Global Communications. There is hereby established within the White House Office an Office of Global Communications (the "Office") to be headed by a Deputy Assistant to the President for Global Communications.

Sec. 2. Mission. The mission of the Office shall be to advise the President, the heads of appropriate offices within the Executive Office of the President, and the heads of executive departments and agencies (agencies) on utilization of the most effective means for the United States Government to ensure consistency in messages that will promote the interests of the United States abroad, prevent misunderstanding, build support for and among coalition partners of the United States, and inform international audiences. The Office shall provide such advice on activities in which the role of the United States Government is apparent or publicly acknowledged.

Sec. 3. Functions. In carrying out its mission:

(a) The Office shall assess the methods and strategies used by the United States Government (other than special activities as defined in Executive Order 12333 of December 4, 1981) to deliver information to audiences abroad. The Office shall coordinate the formulation among appropriate agencies of messages that reflect the strategic communications framework and priorities of the United States, and shall facilitate the development of a strategy among the appropriate agencies to effectively communicate such messages.

(b) The Office shall work with the policy and communications offices of agencies in developing a strategy for disseminating truthful, accurate, and effective messages about the United States, its Government and policies, and the American people and culture. The Office may, after

---

19. John T. Woolley and Gerhard Peters, *The American Presidency Project* [online] (Santa Barbara, CA: University of California (hosted), Gerhard Peters (database)). http://www.presidency.ucsb.edu/ws/?pid=61379 (accessed 1 July 2008).

consulting with the Department of State and obtaining the approval of the Assistant to the President for National Security Affairs on the President's behalf, work with cooperating foreign governments in the development of the strategy. In performing its work, the Office shall coordinate closely and regularly with the Assistant to the President for National Security Affairs, or the Assistant's designee.

(c) The Office shall work with appropriate agencies to coordinate the creation of temporary teams of communicators for short-term placement in areas of high global interest and media attention as determined by the Office. Team members shall include personnel from agencies to the extent permitted by law and subject to the availability of personnel. In performing its functions, each information team shall work to disseminate accurate and timely information about topics of interest to the on-site news media, and assist media personnel in obtaining access to information, individuals, and events that reinforce the strategic communications objectives of the United States and its allies. The Office shall coordinate when and where information teams should be deployed; provided, however, no information team shall be deployed abroad without prior consultation with the Department of State and the Department of Defense, and prior notification to the Office of the Assistant to the President for National Security Affairs.

(d) The Office shall encourage the use of state-of-the-art media and technology and shall advise the United States Government of events, technologies, and other communications tools that may be available for use in conveying information.

Sec. 4. Administration. The Office of Administration within the Executive Office of the President shall provide the Office with administrative and related support, to the extent permitted by law and subject to the availability of appropriations, as directed by the Chief of Staff to the President to carry out the provisions of this order.

Sec. 5. Relationship to Other Interagency Coordinating Mechanisms. Presidential direction regarding National Security Council related mechanisms for coordination of national security policy shall apply with respect to the Office in the same manner as it applies with respect to other elements of the White House Office. Nothing in this order shall be construed to impair or otherwise affect any function assigned by law or by the President to the National Security Council or to the Assistant to the President for National Security Affairs.

Sec. 6. Continuing Authorities. This order does not alter the existing authorities of any agency. Agencies shall assist the Deputy Assistant to the President for Global Communications, to the extent consistent with applicable law and direction of the President, and to the extent such assistance is consistent with national security objectives and with the mission of such agencies, in carrying out the Office's mission.

Sec. 7. General Provisions.

(a) This order is not intended to, and does not, create any right or benefit, substantive or procedural, enforceable at law or equity by any party against the United States, its agencies, instrumentalities or entities, its officers or employees, or any other person.

(b) Nothing in this order shall be construed to grant to the Office any authority to issue direction to agencies, officers, or employees.

GEORGE W. BUSH
The White House,
January 21, 2003.

[Filed with the Office of the Federal Register, 8:45 a.m., January 23, 2003.]

Appendix B

Chronology

20 April 1898	Congress retroactively declares war on Spain.
10 December 1898	Treaty of Paris formalized end of Spanish-American War.
4 February 1899	Philippine-American War begins.
Spring 1916	Major Douglas MacArthur becomes the Army's first public affairs officer.
2 April 1917	President Woodrow Wilson requests Congress approve a Declaration of War on Germany.
13 April 1917	Establishment of Committee on Public Information (Creel Committee).
18 May 1917	President Wilson signs Selective Service Act into law.
5 June 1917	Selective Service registration day (nearly 10 million men presented themselves to enroll in the draft).*
15 June 1917	Espionage Act passed.
8 January 1918	President Wilson's Fourteen Points Speech.
16 May 1918	Sedition Act passed [repealed in 1921].
8 February 1918–13 June 1919	Publication of *The Stars and Stripes.*
September 1939	Office of Government Reports established.
16 September 1940	Selective Service Act signed into law.
16 October 1940	Selective Service Registration Day.
6 January 1941	Roosevelt's Four Freedoms Speech.
March 1941	Division of Information of the Office of Emergency Management established.

---

*David M. Kennedy, *Over Here: The First World War and American Society* (New York, NY: Oxford University Press, 1980), 149–150, 154.

July 1941	Office of Coordinator of Information established.
August 1941	Foreign Information Service established.
14 August 1941	Promulgation of the Atlantic Charter as agreed to by President Roosevelt and Prime Minister Churchill.
October 1941	Office of Facts and Figures established.
December 1941	Office of Censorship established.
5 April 1942	The first broadcast of *Army Hour*.
13 June 1942	Office of War Information established.
1946	Army Information School established at Carlisle Barracks.
January 1948	Smith-Mundt Act passed.
17 March 1949	Secretary of Defense Forrestal creates the Office of Public Information (OPI) in the Department of Defense.
April 1951	President Truman established the Psychological Strategy Board.
1951	Army Information School renamed Armed Forces Information School.
24 January 1953	President Eisenhower forms the President's Committee on International Information Activities.
March 1955	Planning Coordination Group established.
1964	Armed Forces Information School renamed Defense Information School.
1969	Office of Communications set up in the White House.
25 October 1983	US launched operation URGENT FURY, an invasion of Grenada.
1991	Gulf War Military-Media Hearings.
1999	Presidential Decision Directive 68, International Public Information.
2003	Office of Global Communications set up in the White House.

Appendix C

US Government Communication/ Information Departments and Agencies

Central Intelligence Agency. Established under the 1947 National Security Act.

Committee on Public Information. Established by President Woodrow Wilson in 1917. Referred to as the Creel Committee, named after its director, George Creel.

Department of the Army, Office of Information.

Department of Defense, Office of Public Information. Established by Secretary of Defense James Forrestal on 17 March 1949.

Department of State, Bureau for Public Affairs.

Division of Information. Established in March 1941. Robert Horton served as the first director.

Foreign Information Service. Established in August 1941. It was a subdivision of the Office of Coordinator of Information. Robert Sherwood served as the first director of the Foreign Information Service.

Military Assistance Command Vietnam, Office of Information. Established spring 1964. Barry Zorthian served as the first director.

Navy Department, Public Affairs.

Office of Censorship. Authorized under the first War Powers Act of 1941. President Roosevelt appointed Byron Price as the first director on 19 December 1941.

Office of Coordinator of Information. Established by President Roosevelt in July 1941. William G. "Wild Bill" Donovan served as the first director.

Office of Facts and Figures. Established by President Roosevelt in October 1941. Archibald MacLeish served as the first director.

Office of Global Communications. Established by President Bush in July 2002, and made official by Executive order on 21 January 2003.

Office of Government Reports. Established by President Roosevelt in September 1939. Lowell Mellett served as the first director.

Office of War Information. Established by President Roosevelt in June 1942. Elmer Davis served as the first director. The Office of War Information absorbed the functions of the Office of Facts and Figures, the Office of Government Reports, and numerous aspects of the Office of Coordinator of Information and the Foreign Information Service.

Psychological Strategy Board. Established by President Truman in April 1951. It was dissolved by President Eisenhower in the fall of 1953, and its functions transferred to the Operations Coordinating Board.

US Information Agency. Established in 1953. In 1978 its functions were consolidated into the Department of State, Bureau of Educational and Cultural Affairs.

Voice of America. Established in February 1942 as part of the Office of War Information.

War Department, Public Affairs.

About the Author

Mr. Robert T. Davis II joined the Combat Studies Institute (CSI) in June 2007. He earned a B.A. in History from the University of Kansas in 1998, and his M.A. and Ph.D. in Modern European History from Ohio University in 2003 and 2008, respectively. He is the author of CSI Occasional Paper 27, *The Challenge of Adaptation: The US Army in the Aftermath of Conflict, 1953–2000*. Dr. Davis is currently researching the development of the hybrid warfare concept and preparing a manuscript for publication on NATO Strategy during the Cold War.

GPO US GOVERNMENT PRINTING OFFICE: 2009—553-575